Environmental Issues in Vehicle Design and Manufacturing

SP-1579

GLOBAL MOBILITY *DATABASE*

All SAE papers, standards, and selected books are abstracted and indexed in the Global Mobility Database

Published by:
Society of Automotive Engineers, Inc.
400 Commonwealth Drive
Warrendale, PA 15096-0001
USA
Phone: (724) 776-4841
Fax: (724) 776-5760
March 2001

2002 09 24

ISBN 0-7680-0726-7
SAE/SP-01/1579
Library of Congress Catalog Card Number: N98-42939
Copyright © 2001 Society of Automotive Engineers, Inc.

PREFACE

Environmental Issues in Vehicle Design and Manufacturing (SP-1579) contains papers presented at the following sessions of the SAE 2001 World Congress:

- Design for the Environment
- Life Cycle
- Vehicle Recycling Session
- Regulatory and Policy Session, and
- Design and Manufacture for the Environment

These sessions were planned by the Environmental Sustainability Standing Committee and demonstrate the advances in transportation to meet the challenges of sustainable development into the future. It is well understood that both transportation and the environment are essential aspects to the quality of life that we enjoy in a modern world. Our life would be much worse if either were compromised. The authors of the papers herein are working to improve both the environment and the transportation industries. Their work serves as a solid foundation for building industries that protect our future.

This volume would not be possible without the session organizers of the Environmental Symposia at the 2001 SAE World Congress:

James Anderson
Ford Motor Company

Doris Hill
General Motors Corporation

Angelika Coyle
Delphi Automotive Systems

Dennis Havlin
Ford Motor Company

Elisa Cobas-Flores
Environmental Management Engineering

Panel Organizers

Charlette Geffen
Pacific Northwest National Labs

Andrew Haber
NRI Industries

Monica Prokopyshen
DaimlerChrysler

Session Organizers

These people gave unselfishly to seeing that the papers were submitted, edited and finalized for this publication.

Walter W. Olson, Ph.D, P.E.
University of Toledo
Central Technical Committee Chair

TABLE OF CONTENTS

Supply Chain Management: Responding to 'Offsite' Environmental Management

Steven B. Young, Kevin Brady and Andrea J. Russell
Five Winds International

ABSTRACT

Product manufacturers are placing new and detailed environmental requirements on their suppliers. These supply-chain management (SCM) initiatives include both compulsory and voluntary requirements, such as requirements for Environmental Management Systems (EMS), Design-for-Environment (DfE) programs, restricted material conformant, take-back commitments, or performance disclosures. In effect, environmental management is no longer limited to site-specific concerns of emissions and waste. Newer product-focused issues now cover issues beyond the site, from product composition, materials selection, recyclability, product-use, to product take-back. In addition, supplier verification of requirements is becoming an important consideration. These trends are illustrated based on experiences and databases with existing company programs in automotive, electronics, telecom, consumer goods and other sectors. A framework for SCM environmental information is outlined, with an eye to demonstrating opportunities and value-added.

INTRODUCTION

Product manufacturer's are placing new and detailed environmental requirements on their suppliers. These supply-chain management (SCM) initiatives include both compulsory and voluntary requirements, like Environmental Management Systems (EMS), Design-for-Environment (DfE) programs, restricted material lists, take-back commitments, and performance disclosures.

Drivers for these requirements are both internal and external. Internally, environmental management systems, like ISO 14001, sometimes include 1) plans to address environmental impacts and aspects of products and services, or 2) procedures and requirements related to suppliers and contractors. External drivers include government pressures (regulatory, voluntary, and economic mechanisms), end-customer requests, and pressures from competitors in the market. Ultimately, DfE and supply-chain management programs are implemented for business reasons. This includes

possibilities for increased revenue, but also concerns reductions in cost and in product liabilities.

In effect, environmental management is no longer limited to site-specific concerns of emissions and waste. Newer, product-focused issues now encompass "off-site" considerations like product composition, materials selection, recyclability, product-use, and product take-back.

Experience has results revealed a variety of mechanisms, programs and procedures in use across different companies and sectors. A comprehensive supplier environmental requirement may demand both unique capabilities of suppliers and new reporting disclosures. These include: EMS; product DfE practices; lists of banned and restricted substances; process, manufacturing, and pollution prevention requirements; requirements for future product take-back; product information reports; declarations of materials and chemicals; supplier self assessments, evaluations, scoring systems; and analysis of transportation of goods.

Moreover, to increase business value, product manufacturers are using environmental criteria in their purchasing and materials selection procedures to reduce their costs and product liabilities:

1. To integrate environmental criteria into product development and design. For example, through DfE checklists or design guides.

2. To manage environmental attributes of purchased components, engineering specifications and requirements for suppliers. For example, through "Supplier Environmental Questionnaires."

3. To ban or control the presence of undesired substances in products and components as part of a DfE program. For example, "Restricted Materials Lists."

These activities are integrated into engineering, design, purchasing and other business functions across organizations. All together, these trends indicate that life-cycle management approaches, like DfE and SCM, have emerged as a powerful and growing framework for environmental management of products and the supply chain.

SUPPLY-CHAIN MANAGEMENT & ENVIRONMENTAL REQUIREMENTS

Supply-chain management (SCM) is the organization of activities to address the performance of material, components, goods and services that an organization buys and uses. Our research discovered there is a trend toward increasing environmental demands down the supply chain from product manufacturers.

Product manufacturers are requiring their suppliers to undertake similar environmental initiatives that they have such as: ISO 14001, DfE programs, take-back commitments, and Life Cycle Assessment (LCA) studies. Often the focus is on product-transactions in vendor-buyer relationships. Stricter product requirements generate incentives for suppliers to become more innovative and, some argue, can generate value within supplier companies. These initiatives are referred to generically as environmental supply-chain management, referring to the dynamics of the purchasing relationship between buyer and supplier. SCM results in a form of "off-site environmental management", where manufacturers are able to go beyond the traditional boundaries of their site, to manage the environmental aspects of their products and activities.

There are numerous examples of supply-chain environmental management. The ground for many of these initiatives was broken in the automotive industry over the last decade. In Smyrna, Tennessee, at Nissan's North American Operations, plant managers decided to instill in their suppliers a greater accountability for their own packaging waste in the early 1990's. They did so by offering their suppliers long-term, secure contracts that gave them greater stability and predictable revenues and returns. The plant managers also worked very closely with their suppliers to ensure they understood the environmental and economical benefits of using reusable containers and reducing wastes. By the mid 1990s, over 97% of 9750 different parts Nissan shipped into the Smyrna plants arrive in reusable containers (McInerny and White, 1995). Partly as a result of this supply-chain environmental management initiative, Nissan decreased the amount of parts in its inventory by 40% in two years and suppliers, in turn, received secure income, guaranteed business and environmental management training.

More recently, General Motors released a statement (GM, 1998) to its suppliers that placed new environmental management demands on them. The letter requires suppliers worldwide to:

1. Implement and maintain an environmental management system that is consistent with an external standard (like ISO 14001) or a model appropriate for their businesses and locations.

2. Help GM understand how it can best identify, promote and support ongoing environmental performance improvement in its supply chain, and promote similar efforts with next-tier suppliers.

The main objective of the letter is to 'realize ongoing improvement in environmental performance in the GM supply chain.' In doing so, GM feels that the overall value of their products will increase due in part to their decreased impact on the environment.

Ericsson (1998) initiated a comprehensive SCM program; their approach serves as an excellent illustration of tools and techniques that are employed. The firm has basically developed and implemented a set of Supplier Environmental Requirements based on the product life cycle approach, and consistent with LCA:

- suppliers must take a DfE approach when selecting materials or new designs

- suppliers must be prepared to supply LCI data or LCA studies for their products and processes on request

- suppliers are required to abide by Ericsson's banned substances requirements and shall avoid using Ericsson restricted substances

Ericsson's SCM program was developed to help address some of their five main goals that drive their environmental activities. These goals include:

- Development of a common database system for LCA;

- Creation of routines for environmentally adapted design;

- Implementation of environmental management systems;

- Implementation of computer support for materials declaration; and,

- Establishment of recovery and scrapping methods.

The SCM requirements help Ericsson to collect data for their LCA work, to encourage the incorporation of environmental factors into the design of products they purchase from suppliers (thereby improving and adding value to their own products), and in general to reduce their burden on the environment both upstream and downstream.

PRACTICAL EXPERIENCES IN COMPANIES

METHOD - This research report is a composite of recent consulting experience and client-driven studies. Specifically on supply-chain management and DfE, Five Winds International L.P. has surveyed end-use sector experts in more than fifty companies. This includes product manufacturers and suppliers to major product manufacturers in North America, Europe and Asia. A variety of modes of communication were utilized, including telephone, email and personal interviews. Interviewees included company managers, engineers, designers and other decision-makers familiar with product development, manufacture, environmental management, and supplier management.

Although some results are cited by company-name, it was often necessary to survey experts on a confidential basis. Consequently, some results (like company policies and specific guidelines) are cited; whereas others are provided as part of the general analysis, in a manner that shields their origin.

Excellent responses were successfully obtained from more than thirty companies. Sectors covered included automotive (manufacturers and parts), electronics, telecom, building and construction, military and aerospace, and consumer products. The research provided a spectrum of information. This varied from open discussions and sharing of various relevant companies' documents on "environmental supplier programs," "restricted materials lists" and "design for environment," to confidential and very limited exchanges in interviews.

SECTOR FINDINGS - Automotive - All automobile product manufacturers surveyed have developed programs that include elements of SCM and DfE. For example, environmental materials selection was used in all responding firms. This is done typically as part of design guidelines and usually includes banned and restricted lists as well as environmental requirements for suppliers. LCA studies have been undertaken or commissioned by most carmakers, particularly in Europe. Auto companies are responding, in turn, and are passing requirements onto parts suppliers and materials producers.

Volvo AB and the German car manufacturers have taken leading positions with respect to environmental management. Whereas Volvo's environmental programs are rather independent, the German product manufacturers have developed some standardized approaches through the German car manufacturers association. They have used this group approach to

influence and dominate suppliers in Europe over automakers based in other countries (e.g. France, Italy, Spain). In particular, both VW and Volvo see themselves as leading with regards to environmental responsibility and performance. The European automobile manufacturers association is now, under the lead from the Germans, undergoing deliberations on European standards for environmental initiatives.

The German industry is working on the harmonization of DfE standards and SCM specifications for a number of issues:

- provision of materials environmental information by suppliers (compositions, recyclability, etc.)

- supply of life cycle information on products in support of LCA studies

- take back obligations for suppliers

- recycling standards

- material lists (bans and restricted lists) plus nomenclature lists for component labeling.

Volvo's activity with regards to environmental materials selection is significant. The company was first to assemble comprehensive environmental requirements for suppliers and to develop "black" and "gray" lists of banned and restricted substances. It was also the first to work rigorously along the supply chain and with competitive companies in the industry. Volvo's influence is very apparent both within the sector and in allied sectors like electronics and electric components. Volvo will likely continue to set much of the standard and expectations in the sector on issues of environmental performance.

The Big 3 North American automobile makers also have programs in place to help determine materials selection on environmental grounds. These initiatives tend to be globally oriented. Moreover, the USA-based automakers have recently completed a comprehensive total-vehicle life-cycle inventory study, with the assistance of the aluminum, steel and plastics industries, as well as part suppliers.

Electronics & Telecom - The electronics and telecom sectors are positioned similarly to the automotive industry. Many of the surveyed companies have in place comprehensive programs like those in the automobile sector; however, the emphasis and terminology is focused more on DfE.

Most electronics and telecom firms have DfE programs, tools or approaches that allow them to assess or screen products for environmental aspects during design. In fact the implementation of DfE, as a more integrated approach to product change compared to simple Restricted Materials Lists is more widespread compared to automotive. These practices almost certainly address issues of packaging reuse and return, but also more and more are focused on primary product production.

The European electronics waste regulation (WEEE 1998) is an especially powerful driver. But the legislation, due to take effect soon, has influenced manufactures around the world – not just in Europe. This activity has lead to European customers asking for more detail on what they are buying from electronics suppliers, primarily for business concerns, as they want to avoid costs under the new waste regulation and to reduce their product end-of-life liability.

Take-back has also stimulated competition within the electronic sector – for example, telecommunication companies are trying to lower the cost of ownership for their customers and thereby gain competitive advantage.

Building and Construction - Findings indicated the building and construction sector has as yet to develop beyond its current minor capacity. It was observed that there is an emerging interest in environmental materials selection. ABB, the large heavy industry and power engineering firm, has a significant environmental program centered on environmental management systems and services; although, the company offered only a little detail regarding environmental materials selection.

Generally, this sector is expected to grow its environmental activities. Numerous initiatives are underway to use environmental criteria in design and materials selection. Of note, most major flooring systems companies have completed comprehensive product-focused studies (including vinyl flooring, wood and carpet systems). In the residential housing sector, there is heated competition between wood, steel and concrete designs, where environmental arguments address building life-cycle energy, construction waste, building energy-efficiency, and end-of-life disposal.

The SCM model is not as prominent in the building sector. Compared to automotive or electronics, the vertical integration and tier-structure are not apparent. Architects, that are often relatively small firms, or development companies, which tend to be regionally based, appear to be at the top of the decision-ladder. This does not compare to large, integrated and global entities that are found in electronics and automotive.

Emerging for architects, designers and engineers are numerous analytical tools. These are generally LCA-based tools that are data intensive in their analysis of life-cycle impacts associated with building materials, buildings and associated infrastructures like roads and bridges. Other mechanisms for DfE in this sector are conferences, like the Green Building Challenge, and

resource documents like the American Institute of Architects' Environmental Resource Guide [AIA, 1998].

Other Sectors - Companies in other sectors surveyed (consumer products, military and aerospace) do have programs, although it was found that capacity and implementation vary significantly from firm to firm.

First generation supply-chain management programs tend not to include product or life-cycle content in their scope. A number of North American manufacturers, for example, are in recent years initiating some proactive environmental initiatives. The focus tends to be on suppliers' facilities, their practices and their record of regulatory compliance. No product-focused information is requested or expected.

PRACTICAL COMPANY EXAMPLES - Compaq Compaq, the USA-based computer company, implements DfE directly. Its approach focuses on environmental stewardship during every phase of the product life cycle, borrowing heavily from the LCA concept. When Compaq engineers begin the design of a computer, they consider the environmental impact of component parts and their readiness to be recycled when the computer is no longer useful.

Formalized in 1994, DfE Guidelines have been developed for use across Compaq product lines on a worldwide scale. They emphasize several key principles:

Energy Conservation: Products that minimize energy consumption during both active and inactive periods facilitate preservation of worldwide energy resources. A focus on energy conservation reduces customer-operating costs.

Disassembly: Concepts learned from Compaq's global network of recycling partners have been integrated into mechanical design practices. Product designers plan for the end of a product's life, when computer components can be recycled.

Reuse and Recyclability: Compaq products use materials that can be easily identified and recycled, employing available recycling infrastructures. Ease of recyclability will help its customers reuse or recycle products at the end of their useful lives.

Packaging: Packaging materials are the first product aspect a customer sees and handles. Additionally, packaging is the initial source of waste generated by a product once it enters the market. Kept to a minimum under DfE Guidelines, packaging is composed of recycled materials and uses no heavy metal inks. The company requires at least 35 % recycled materials in all corrugated papers for packaging. All of Compaq's packaging is made of materials that can be easily identified and recycled.

Upgradability: Compaq incorporates design features that will aid in extending the life of a computer. Features such as microprocessors, memory, internal storage and other subsystems that can be upgraded will aid in the prevention of early product obsolescence.

Compaq integrates numerous features into the design and manufacture of its products in an effort to reduce environmental impact over the life cycle. In Europe, two recognized eco-labels signify the company's dedication to manufacturing environmentally-sound products. In Germany, Compaq uses the Blue Angel eco-label and in Sweden it is the TCO '95 designation. These programs identify products that are environmentally friendly, recyclable and easily serviced. Compaq's commercial desktop product families are certified with the German Blue Angel eco-label. The Blue Angel indicates these Compaq desktop computers minimize negative impact on the environment in their design, ergonomic properties, safety, power consumption and documentation. The complete range of Compaq commercial monitors are certified for the TCO '95 emblem, an indicator of monitors that are energy-efficient and emit minimal electric and magnetic fields.

In the area of SCM, Compaq has developed a supplier development process to regulate their supply chain. The suppliers are given a questionnaire to fill in information about their policy and commitment, compliance history, processes and assessment, their suppliers, CFCs and other hazardous materials, and their waste minimization initiatives. In terms of material specification, the suppliers' policies must meet or exceed regulations. Compaq's environment health and safety criteria are used to track compliance on parts and purchasing contracts.

Chrysler - At Chrysler "Life Cycle Management" is the name for its environmental SCM approach. However, most importantly from a business perspective, the program is largely driven by economics, not environment. The firm estimates hidden costs of restricted substances to be $2.5 billion over five years. The feeling is that this cost could be all translated into profits associated with vehicle launch and sales. The initiative provides Chrysler with a good overview on the relevant environmental issues of their products. Their strategy is to comply with the law plus avoid non-value added costs.

Costs are determined via analysis of environmental and regulatory information: abatement cost, waste treatment, emissions control, filters, off-site hazardous disposal, administration, liability, training, medical testing, lost opportunities in recycling, need for over design, etc. The approach enables Chrysler to assess the various products and processes, both in-house and at the supplier – and enables fast decision support in the design stage. Chrysler is forcing requirements onto its suppliers, while simultaneously expecting them to reduce costs to Chrysler.

Extended material ban lists are driven by cost criteria (abatement and health/safety) and regulatory issues. The bases for the selection of targeted substances is via analysis of regulations, looking at international, national, regional and local requirements. They also examine bellwether jurisdictions to get a feel for the future direction of regulation. In the US, the conventional lists are covered: Clean Air Act, TRI and other current US legislation. They have developed a hierarchy of substances through analysis – looking at carcinogenicity and teratogenicity data from Sweden, Germany, Netherlands, Ohio, Wisconsin, Indiana, Minnesota, Michigan, and EPA. They identified 1700 substances and found that Chrysler actually used 761 of these. Of these, 103 compounds represent 90% of Chrysler's risk and reporting cost.

Siemens NAMO (North American Motor Operations) - Siemens NAMO produces electric motors for automotive applications, from a plant in London, Ontario, Canada. Volvo is a major customer and, consequently, has influenced the Siemens' environmental program significantly. Both SCM and DfE approaches have been implemented.

The overall corporate focus at Siemens NAMO has been on eco-efficiency for many years. In 1993, a corporate environmental officer was quoted as saying: "Economy without ecology is irresponsible, ecology without economy is naïve". Part of their Corporate Environmental Policy statement reads: 'If we let a coherent ecological analysis inform the way we think and act, this almost invariably produces economic benefits.' These statements indicate the type of corporate culture Siemens NAMO has established and maintained over the last decade of their environmental activities.

They use a highly integrated approach to production at their facilities; minimizing the use of resources and the production of waste wherever possible. Furthermore, they have integrated environmental decisions/concerns into their product development process, requiring their designers to use DfE checklists and guidelines for restricted and preferred substances. Siemens also uses LCA for product research and to help inspire innovation in the company. Their approach to product design can be explained by the following quote from their 1998 Environmental Report:

'Environmentally-sensitive product design, which covers the entire life-cycle of an item of equipment or an installation, is a classic example of economic benefits resulting from environmental concerns'.

EMS at Siemens NAMO is part of a larger "environmental stewardship" program that combines LCA with EMS. After drafting their ISO 14001 audit protocol in

1996, Siemens then embarked upon a two-year certification implementation plan. Four of their facilities were selected to apply for certification to ISO 14001 in 1998. The EMS work on aspects/impacts was also extended to use concepts of inputs and outputs, derived from the ISO 14040 standard on LCA

An obvious key benefit of implementing an EMS is the short term cost savings derived from the aspect/impact evaluations. In 1998, the London plant achieved savings of $600,000. These consisted of recycling reduction and recycling of metals, paper/cardboard, pallets, gloves, cloths, component reuse and delisting of wastes. Landfill costs were reduced in 1998 by 95% over 1994 costs and recycling rate reached 85.2%. Hazardous waste disposal costs were reduced by 75% over 1994 costs.

Siemens NAMO developed a SCM Audit protocol, which helps insure compliance to the environmental goals and programs of customers in the Automotive Division, especially those dealing with banned substances. Each of the 100 or more parts suppliers to the London facility has received a self-audit program. This package contains a points oriented audit and Black and Grey lists. Substances on the Black list are not to be part of the component, or the process used to manufacture the part. Substances on the Grey list must be identified and a timely plan developed to eliminate them.

The data collected from the aspects/impacts study was used in the development of plant wide environmental goals that identified opportunities to reduce waste and unnecessary materials within the production systems. The next objective is to develop a fully integrated system for all of Siemens' production processes, which will minimize unnecessary aspects that add cost to the manufacturing process. All incremental costs can be accurately ascribed to each product, including energy requirements.

The principles of ISO 14040 standard for LCA and DfE were first used in 1997, with the redesign of Siemens NAMO brushless motor. This motor lasts 10,000 hours compared to 2000 for the old design and is assembled rather than welded, generating lower cost manufacture and ease of disassembly. Moreover, the new design has a life expectancy improvement of 80% over a regular fractional horsepower motor and is designed for disassembly.

With the development of the high efficiency cooling fan motors, Siemens NAMO is currently developing modules to replace the drives from the engine for both the water pump and the power steering. On a 1200 kg automobile traveling at 90 km/h, an 8% fuel saving is conservatively achievable.

Home Depot and Forest Products suppliers - As a solid step in its SCM, The Home Depot established its Timber Task Force in 1998 to improve worldwide forest practices

and review the environmental practices of key vendors. Key vendors include forest products companies such as Weyerhaeuser and Canfor. The Task Force conducts research into sustainable forestry initiatives and the company has helped to pioneer a marketplace for wood products certified by the Forest Stewardship Council of the Certified Forest Products Council (CFPC), of which the company is a member. In 1999, Annette Verschuren, president of The Home Depot Canada and chairwoman of the group's environmental council (Home Depot's membership at a British Columbia Forest Industry Conference in Vancouver) Annette Verschuren, stated:

> *"We intend to work closely with our vendor partners to develop products that our customers want, need and deserve -environmentally preferable wood products from certified well-managed forests. We challenge other home improvement retailers and mass merchandisers to join us."*

The company lists products on its website that have been independently certified by the third party independent FSC. In the last six years, they have introduced numerous certified products for their customers. In addition, the company communicates externally its activities in this area and provides descriptions of many products that are recycled, efficient, or otherwise environmentally notable. The firm promotes the use of alternative materials and products to its customers and requires manufacturers to submit products to a third party verifier for evaluation of the accuracy of any environmentally preferable product claims.

In the area of eco-labeling, Home Depot was the first home improvement retailer in the world to promote and use "Environmental Report Cards" or to offer "eco-profiles" of products. These eco-profiles are based on studies that assess the environmental burdens of a product over that products' life cycle.

Both in foresight and as a result of Home Depot's requirements, many of their suppliers of forest products are establishing programs to collect LCA data and implement sustainable forest management practices. By doing so, these companies not only help maintain contracts with companies like Home Depot, but can improve their environmental practices.

Canfor - One of Home Depot's suppliers is Canfor, a Canadian-based forest products company, producer of both wood and pulp. Canfor recently sponsored and participated in a joint LCA study of graphic paper and print products (magazines and newspapers) to determine the most significant environmental aspects for these product systems.

In a separate initiative, Canfor produces third-party certified Environmental Product Declaration Sheets

(EPDS) to provide its customers with environmental profiles of their products and is currently implementing ISO 14001 at the majority of its facilities. EPDS is a form of Type III labeling, coordinated by the Canadian Pulp and Paper Association. Canfor also participates in the Sustainable Forestry Initiative with the USA-based American Forest Products Association.

Weyerhaeuser - Weyerhaeuser, another supplier to Home Depot, has produced environmental profile fact sheets for many of its products and its operations for both its customers and stakeholders.

Weyerhaeuser is also very involved in the Sustainable Forestry Initiative with the American Forest Products Association - this helps to secure their contracts with customers such as Home Depot.

To help maintain its preferred supplier status, Weyerhaeuser is pursuing ISO 14001 Certification at the majority of its facilities, including forest operations. Doris Zacher, Director of Information and Data Management for the Corporate Environment, Health, and Safety Group, had the following to say about the benefits of ISO 14001 implementation:

"Generally, we think that ISO 14001 will help protect our preferred supplier relationships with key customers. There are a lot of stakeholders who are interested in our EMS capability, including state and provincial regulatory agencies. The effect of having an ISO 14001 EMS will be positive in the future...It will also help us meet customer and stakeholder expectations. The ISO 14001 standard is solid and internationally recognized. This framework allows Weyerhaeuser to deploy a common approach to manufacturing operations in the United States, Canada, and offshore."

DISCUSSION

Many product manufacturers are realizing that environmental SCM is an important part of their overall supply-chain management approach. To monitor and manage the environmental performance and management systems of suppliers has become commonplace, with differing degrees of sophistication. SCM allows a purchaser to standardize expectations, requirements and practices of their suppliers – possibly including supplier facility environmental management, product design approaches, composition of supplied parts and components, and provision or disclosure of data.

SCM exists in many companies without environmental content, that is SCM is "supply chain management" – where the focus is on performance, cost, quality and timely delivery. In the context of the present discussion,

"environment" is consciously added to the SCM list of criteria. Depending on the organization, environmental SCM will bring in a variety of programs (like auditing), techniques (like surveys and checklists) and possibly impose systems requirements on suppliers (like ISO 14001 EMS). One of the approaches that is often relevant is DfE.

From another perspective, DfE is the dominant approach. In this model, environmental criteria are integrated into the design process. Techniques (like checklists and restricted substance lists) and programs (like SCM) then become appropriate, either in the design process itself or in the implementation of product changes. And, clearly, to affect comprehensive changes to products, as might be directed by DfE, there are many actions required. Implementation of an EMS is one outcome of DfE, as are changes to transport or manufacturing. If activities to be changed are in the hands of suppliers, then an environmental SCM program is one obvious and valuable route to implement DfE.

From any perspective, it is certain that one trend can be anticipated, and, in fact, has already begun occurring; OEMs asking suppliers for Life Cycle Inventory profiles of their products. Initially this was cause for concern for suppliers because it meant handing over highly confidential data containing detailed process information. However, there are now tools available that can protect suppliers from handing over confidential data, while still allowing OEMs to obtain the information they require for their DfE or other environmental programs. These tools are structured software programs designed to manage life-cycle data, while respecting information sensitivity. As long as the suppliers are reporting data to the OEM using the same software program, this protection of confidential data is possible. The data can be coordinated and reported in several scenarios, depending on the complexity of the supply chain. Figure 1 indicates different levels of data use and containment; corresponding to tiers in the supply-chain. These different levels show how data flow can be coordinated and protected along the supply chain. In product-environmental management circles, these packets of data are analogous to "environmental product declarations." In general, this type of information management software allows companies to gain a comprehensive understanding of the environmental impacts of their products without putting suppliers in a compromising and uncomfortable business situation. Note, however, that the basic building blocks for product environmental data (materials, energy, etc.) are public and need to be verified by all users from lowest tier to the OEM.

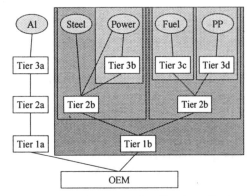

Figure 1. Proposed Routes for Product Environmental Information Flow in the Supply Chain

CONCLUSION

Our research and experience has found that a variety of mechanisms, programs and procedures are in use for SCM across sectors and in different companies within those sectors. To increase business value, reduce risk, and manage liability, many product manufacturers are incorporating environmental requirements into their procurement practices. This is being done in several ways, including:

- "DfE Programs" - the integration of environmental criteria into product development and design.

- "Supplier Environmental Questionnaires" - attempts to manage the environmental attributes of purchased components, engineering specifications and requirements for suppliers.

- "Restricted Materials Lists" - eliminating or monitoring the presence of undesired substances in products and components.

These activities are integrated into business functions such as engineering, design, and purchasing, across organizations. In general, these trends indicate that life-cycle management approaches, like DfE and SCM, have emerged as a powerful and growing framework for environmental management of products and the corporate supply chain.

REFERENCES

1. American Institute of Architects, Environmental Resource Guide, AIA: NY, NY, 1997.

2. R. Barton, L. Rowledge, K. Brady, Mapping the Journey, Greenleaf Publishers: UK, 1999.

3. Canfor, personal communications and company literature, 1998-2000.

4. Chrysler, personal communications and company literature, 1998-2000.

5. Compaq, personal communications and company literature, 1998-2000.

6. Ericsson, www.ericsson.se, accessed Jan. 2000.

7. Fava, J., et al, A flexible framework to select and implement environmental strategies, Journal for Strategic Environmental Management: Volume 1, Issue 1, 1998.

8. General Motors, (1998). Environmental Performance Management in GM's Value Chain. Worldwide Purchasing and North American Production Control & Logistics. Company Statement, July 22nd.

9. The Home Depot , www.homedepot.com, accessed February 2000.

10. IBM, personal communications and company literature, 1998-1999.

11. F. McInerney and S. White. (1995). The Total Quality Corporation. Penguin Books USA Inc.: NY, NY.

12. Siemens NAMO (North American Motor Operations), personal communications and company literature, 1997-2000.

13. The Home Depot, personal communications and company literature, 1998-2000.

14. Volvo AB, personal communications and company literature, 1998-2000.

15. WEEE, European Commission, Waste Unit Proposal for a Directive from Electrical and Electronic Equipment (WEEE), July 27, 1998.

16. Weyerhaeuser, personal communications and company literature, 1999-2000.

CONTACT

Dr. Steven B. Young

Five Winds International

s.young@fivewinds.com

www.fivewinds.com

FMEA-Based Design for Remanufacture Using Automotive-Remanufacturer Data

A. Lam, M. Sherwood and L. H. Shu
Department of Mechanical and Industrial Engineering, University of Toronto

ABSTRACT

This paper describes the development of a Failure Modes and Effects Analysis (FMEA) modified to support design for remanufacture. The results of the waste-stream analysis of an automotive remanufacturer were used for this FMEA. The remanufacturer waste stream was assessed to determine factors that impede the reuse of parts. The use of the modified FMEA allows consideration of factors such as ease of detection and repair of failure, in conjunction with contribution to the waste stream of each failure mode, to develop priorities in design for remanufacture.

INTRODUCTION

The long-term goal of this research is to develop a methodology for designing products that can be more easily remanufactured. Remanufacturers take back, disassemble and clean used products, replace or repair failed parts, and reassemble products in a production-batch process. The essential goal of remanufacture is to reuse parts. Parts that cannot be reused are discarded and enter the remanufacturer's waste stream. Therefore, analysis of this waste stream will identify barriers to the reuse of parts. The results of such an analysis may be used to develop design strategies, such as guidelines and metrics, which enable more efficient remanufacture of future products. Specifically in this study, we quantified through data, the FMEA indices of occurrence (OCC), detectability (DET), and repairability (REP), which is our measure of severity (SEV), of the failure modes identified in the remanufacturer's waste stream.

BACKGROUND AND MOTIVATION

Traditionally, products have been designed and manufactured to meet functional needs during the product's useful lifetime, with little regard for the product's end-of-life. Recently, in response to stricter environmental legislation, particularly in Europe and parts of Asia, that assigns responsibility for products at the end-of-life to manufacturers, more products have been designed for ease of scrap-material recycling. Scrap-material recycling involves separating a product into different materials and reprocessing the materials for use in similar or degraded applications.

For appropriate products, remanufacturing offers significant environmental and economic benefits over scrap-material recycling. Remanufacturing involves recycling at the parts level as opposed to the material level. Recycling at the higher level of components avoids resource consumption for possibly unnecessary reprocessing of material. Remanufacturing also postpones the repeated degradation of the raw material through contamination and molecular breakdown, frequently characteristic of scrap-material recycling. Furthermore, remanufacturing can divert parts that are not recyclable for material content from landfill and incineration. The production-batch nature of the remanufacturing process enables it to salvage functionally failed but repairable products that are discarded due to high labor costs associated with individual repair.

The relationship between manufacturing and remanufacturing is depicted in Figure 1, which shows that while the manufacturing process produces new products, the remanufacturing process can repeatedly take products at the end-of-life and transform them to a "like-new" condition for reuse. A few manufacturers remanufacture their own products. In the office equipment market, companies such as the Xerox Corporation have achieved $200 million in annual savings by remanufacturing their photocopiers [U.S. Congress 1992]. Some manufacturers, such as the Ford Motor Company have "Authorized Remanufacturers" to process after-market parts for their cars. With increasing international product take-back legislation, more manufacturers are likely to become interested in the remanufacture of their products.

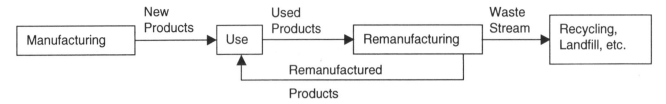

Figure 1. Relationship between manufacturing and remanufacturing.

RELATED WORK

Design for recycling is the focus of much research related to product retirement [Ishii and Lee 1996]. It is an area that is mostly complementary to design for remanufacture. Any of the steps of remanufacture, e.g., disassembly, cleaning, etc., treated as a design-for-x method also supports design for remanufacture. Thus some research in design-for-disassembly include maintenance and remanufacture [Zussman et al. 1994]. Other research has developed strategies to aid product reuse [Umeda et al. 1999, Mangun and Thurston 2000]. Shu and Flowers consider reliability and fastening and joining with respect to remanufacture (1998, 1999). Hammond et al. (1998) conducted a survey of automotive remanufacturers to uncover process difficulties and generated design-for-remanufacture guidelines and metrics [Hammond and Bras 1996]. Sherwood and Shu summarized the results of waste-stream analyses of three automotive remanufacturers and proposed the use of a modified FMEA to support design for remanufacture (2000).

Research that addresses FMEA in the context of redesign is also relevant [Eubanks et al. 1996, Kmenta et al. 1999]. Despite the work on specific aspects and general surveys of remanufacture, a systematic study of remanufacturing difficulties is necessary to ensure that no significant issues are overlooked in the development of design-for-remanufacture strategies.

The approach of this work is novel in that the remanufacturer's waste stream and other process data is quantified to uncover design-for-remanufacture strategies.

REMANUFACTURE INDUSTRIES

Lund (1996) compiled a list of 9,903 remanufacturers and identified the most dominant product sectors as automotive, electrical apparatus, tires, and toner cartridges. The automotive sector, with typical products of alternators, starter motors, water pumps, clutches and engines, comprises the highest percentage, 46% of Lund's database population. Next are electrical apparatus (transformers, electrical motors, switch gear)

at 23%, toner cartridges at 14% and retreaded tires at 12%. Other categories comprise 5% of Lund's database of remanufacturers.

This paper continues to address the results of data collection in the automotive sector. Previous work in this sector is described in Sherwood and Shu (2000). Williams and Shu (2000) present results of data collection for toner cartridges.

REMANUFACTURING PROCESS

At the original-equipment remanufacturer (OER) of automotive engines where this study was conducted, the production-batch process proceeds as follows. The received engines are delivered to the disassembly station. In each batch, seven to fifteen engines, depending on the size of the engine, are disassembled. The engines are dismantled and the parts sorted into baskets for cleaning. The cleaning process uses either chemical spray or high-pressure water. The processes for the different parts are as follows.

Block - After a cleaning process that removes grease and other chemicals from engine blocks, the bore diameters are measured using gauges and compared to specifications. For blocks within specification, threads are tapped before the blocks are sent to machining lines. The bores for the crankshafts are checked for straightness and size, and machining is performed as necessary. Next, the seat is milled and the piston housing is bored. Finally, the blocks are washed to remove metal chips accumulated during the machining processes and await delivery to the assembly line.

Cylinder Head - Cylinder heads disassembled from engine blocks proceed to a subsequent station for further disassembly of the springs, rocker arms, valve pins, etc. The aluminum cylinder heads are then washed in a separate machine since they cannot be treated together with iron and steel parts. The cylinder heads are then sand blasted before threads are tapped. Next, guides for the valve pins and valve seats are replaced before seat cutting or milling is performed. Like the engine blocks, cylinder heads are washed to remove excess metal chips accumulated during machining before proceeding to assembly.

Crankshaft - After being cleaned together with camshafts, connecting rods, oil pans and valve covers, crankshafts are delivered to machining lines. The crankshafts are first gauged to check the dimensions of all shafts. When a part satisfies all the specifications, it proceeds to stations for grinding and polishing. Crankshafts are gauged again before being delivered to the assembly line.

Camshaft - The process for camshafts is similar to that for crankshafts. They are cleaned, gauged, ground and polished. However, after cleaning, instead of proceeding to a machining line, they are brought to a station where usually one employee is responsible for all the remaining processes to be performed on a particular camshaft.

Connecting Rod - There is no machining process in use for connecting rods. After cleaning, all connecting rods are delivered to a station where they are sorted according to engine model. Then, all rods from the same engine type will be loaded onto a shaft that serves as an initial measuring tool. If a rod cannot fit onto the shaft, it is scrapped because the attached cap is likely mismatched. All bolts are replaced before the connecting rods are gauged to check dimensions. Finally, connecting rods are weighed and grouped by weight. Depending on the number required by the engine type, four to eight rods with the same weight, within an accepted tolerance, are grouped together to enhance crankshaft performance.

Oil Pan/Valve Cover - Comparatively, the refurbishment of oil pans/valve covers requires simpler processes. After cleaning, accessible dents are removed from the oil pans/valve covers. If the convex side of the dent is inaccessible, the dent cannot be removed and the pan/cover will be scrapped. The oil pans/valve covers are then painted before proceeding to assembly.

Cylinder Sleeve - After cleaning, cylinder sleeves are sand blasted. The bores are then gauged and the seats polished. Lastly, the cylinder sleeves are bored. Special care must be taken during the boring process and consequently a considerable number of sleeves are machined oversized.

FAILURE AND SCRAP CATEGORIES

Typically, a part must satisfy two conditions before entering the waste stream of a remanufacturer. The first condition is that the part has failed, i.e., it can no longer fulfill its intended function. The second condition is that the part is deemed not repairable. Failure mode refers to why the part cannot be reused without repair, e.g., due to presence of crack, excessive wear or corrosion. Many failed parts can be repaired. For example, some cracks may be welded and dents may be removed. Scrap mode refers to why the part was not repaired, typically due to technical limitations, and includes reasons such as no process is available for repair or the repair is not economically justifiable. Figure 2 depicts the relationship between failure and scrap modes. Each part entering the waste stream was counted and described according to both failure and scrap modes. Sound parts that were scrapped were labeled 'no failure,' as described below.

FAILURE MODES

Fifteen failure categories, including the "no failure" category were identified and described as follows.

Bent - This category includes connecting rods, the occasional crankshaft, and very few large pieces such as cylinder heads that are warped.

Burnt - Parts become burnt due to lack of oil in the engine, a result of maintenance failure by the vehicle owner.

Corrosion - This category includes blocks, crankshafts, camshafts, connecting rods, and oil pans that rust when exposed to water vapor, and cylinder heads that are corroded by cooling agent.

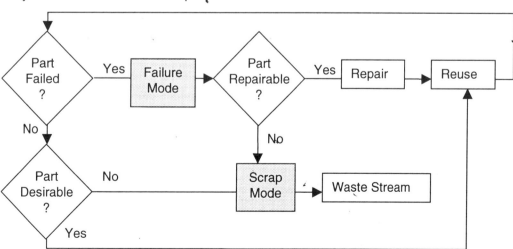

Figure 2. Failure mode versus scrap mode

Crack - Parts that crack include blocks, camshafts, cylinder heads, and cylinder sleeves. Most cracks spotted by disassemblers are deep cracks and tears.

Dent - This category is mostly comprised of oil pans /valve covers.

Design Flaw - Some parts such as blocks have design flaws that are not correctable during remanufacture.

Disassembly Damage - Some parts are damaged during the disassembly process.

Disassembly Impossible - Other parts cannot be reused because they cannot be disassembled.

Fastener Failure - Included here are damaged transmission mounts that project from the sides of the engine block, bolts that snap off in holes and stretched keyways in crankshafts and camshafts.

Fracture - Fractured parts are broken into pieces. While the same types of parts that crack are fractured, cracked parts may be retrievable while fractured parts are typically not repaired.

Handling Damage - These parts are damaged in different ways while being moved.

Hole - Holes are large material segments removed from the part body. Parts with holes are primarily engine blocks and occasionally punctured oil pans.

Machining Damage - Some parts are damaged due to errors in machining.

Wear – Engine-block piston housings, camshaft bearings and cylinder bores are worn during ordinary use, while crankshafts are gouged and grooved from improper engine operation and care.

No Failure - There is a large category of parts discarded that were not functionally flawed, but were scrapped due to the overstock scrap mode detailed below.

SCRAP MODES

Scrap modes correspond to reasons why parts are discarded and correspond to the FMEA category of effect of failure.

Cosmetic - Functional parts may be discarded due to cosmetic imperfections. Examples include plastic parts with scratches and dents that cannot be fully removed.

Last Oversize/Undersize - The repair process for worn, burnt, or corroded parts is further removal of material, sometimes resulting in under- or oversized parts. Parts are at last oversize/undersize for two observed reasons.

First, too much material is lost from the part during use. Second, parts returned to the remanufacturer have had material removed during previous remanufacture processing.

Blocks, cylinder heads, connecting rods and cylinder sleeves have bores that become oversize through wear and/or previous reboring. Crankshafts and camshafts become undersize through material removal. Blocks and crankshafts are the only parts that bear remanufacturers' labels and can thus be more easily identified as having been previously remanufactured.

Makes Oversize/Undersize - This category is similar to "last oversize/undersize" except that the cause of oversize/undersize is within the remanufacturing facility. This occurs when refurbishment requires the removal of an excessive amount of material, such as the case with deep scratches, or the remanufacture worker has erroneously removed too much material in machining. In the first case, the parts would usually be scrapped before refurbishment whereas in the second case, the part is scrapped during or after machining. Again, this scrap mode is mainly associated with wear, burn and corrosion.

The parts scrapped in this category are blocks, cylinder heads, crankshafts, camshafts and cylinder sleeves. Parts requiring boring, such as blocks and cylinder sleeves, are more likely to fall in this category because boring removes more material in one pass than grinding and polishing. In cylinder heads, "makes undersize" usually occurs during seat cutting, which involves the removal of a large amount of material. Even though the depth of cutting is fixed, machining error is still possible. On the other hand, crankshafts are carefully gauged before machining; thus only a few of them are machined undersize.

Material Loss - This category accounts for parts discarded due to material loss that cannot be economically or reliably replaced. Material loss may have occurred through corrosion, scratches, fractures and holes. Patching holes and removing corroded material on blocks are time-consuming and therefore not performed. Cylinder heads lose material around valve seats. This material loss is caused by the impact of valve pins that break off during engine operation. Scratched or fractured camshafts cannot be reliably welded and are scrapped. Other parts scrapped due to material loss, mainly as a result of fracture, include connecting rods, oil pans/valve covers, and cylinder sleeves. Connecting rods may also be discarded because of scratches in the bores. Material loss also has high correlation with the "burnt" failure mode since many of the parts that were scrapped were also burnt.

Mating Part Lost - Some parts, such as bearing caps and connecting rods, wear together and must be kept together. If one piece is missing or mismatched, the part is discarded.

No Process - Parts in this category are scrapped because the remanufacturer does not have equipment or know-how to perform the required repair process. Also included are parts for which the damaged portion is not accessible to the repair process used.

Overstock - Parts are scrapped when new parts can be bought more cheaply than old parts can be refurbished and when there is sufficient inventory and no more storage capacity. Overstock is seasonal and results in a large number of a particular kind of part to be scrapped at once.

Sacrificial Part - This category corresponds to parts that are damaged in disassembly in order to save a more valuable part. A considerable number of connecting rods were scrapped at the disassembly station after being struck out from the block. To salvage the more valuable engine block, the connecting rods are bent, twisted or fractured during removal.

Unknown Damage - This category is designated for those parts that are scrapped without being examined to identify specific damage. For instance, when the remanufacturer receives an engine with a large hole in the block, the entire engine may be scrapped without disassembly, since it is suspected that other severe damage is likely to be present in this engine. Parts that are retained in the engine block and not disassembled are classified in this category, since specific damage was not identified.

Warped Part - This category includes parts that are or are likely to be warped as a result of welding. Parts are frequently preheated before welding. Preheating is done at a higher temperature when welding is required at many locations, or over a large surface, or when the part itself is large. Warping may occur as the part cools. The warped part is scrapped because it cannot be straightened.

Two particular models of cylinder heads had more problems of warping after welding because the cylinder head sizes are larger than other models. Consequently, cracks, scratches and other damages are not repaired on these two types of cylinder heads, resulting in a large number of scrapped parts.

Weakens Part - Some damaged parts can be repaired, but the repair process may adversely affect the function and/or lifetime of the parts. Such parts were scrapped because testing would be too costly. For instance, bent or twisted connecting rods could be straightened, but since the process may weaken the parts, they are scrapped.

DATA COLLECTION

Data collection lasted from May to September 2000 and a total of 1245 parts were counted. The daily production volume of the remanufacturer studied is enormous. Thus, it is not reasonable and feasible to examine every single part that flows through the plant. It was essential to select a certain number of parts for the study. This project was conducted to investigate the occurrence, detectability and the repairability of certain failure modes identified in the waste stream. Therefore, the selected parts must be relevant to these three indices.

The parts studied include blocks, cylinder heads, camshafts, crankshafts, connecting rods, oil pans/valve covers and cylinder sleeves. A route through all the stations on the shop floor was established to effectively collect data. The route started from the disassembly station, through the oil pan/valve cover painting stations located adjacent to the disassembly station, followed by the stations for cylinder heads, connecting rods, crankshafts, blocks, cylinder sleeves and camshafts. The sequence was made according to the shop floor layout.

During each pass, two types of data were recorded. First recorded is the number of parts for which all the required processes at each station had been completed. Next, the scrap bin was inspected and the number of scrapped parts and the discard reasons were recorded.

The remanufacturer studied has contracts with and only processes engines of two OEMs. This characteristic provides stability and consistence of the collected data.

FAILURE AND SCRAP MODE BY PART TYPE

In this section, findings will be discussed on the basis of part type. The dominant failure modes and scrap modes of each part type are described.

Block - The dominant failure mode of blocks is wear of the piston housing bores. This failure mode is associated with the scrap mode of "last oversize". Another possibility is that the bore is machined oversize, corresponding to the categories of "machining damage" and "makes oversize". This possibility is unlikely because all blocks are gauged first and a standard tool is used for boring.

Bores in the blocks for crankshafts are also worn. The crankshaft is fitted into the block with a cap. Wear loosens the cap resulting in the failure mode "wear" and the scrap mode "last oversize."

Another dominant failure mode for blocks is "disassembly impossible". This failure mode occurs when the block cannot be disassembled because connecting rods retained in the block cannot be drawn out by disassemblers. The block is scrapped because the remanufacturer has no process for disassembly.

Cylinder Head - Cracks on cylinder heads may result from the impact of valve pins. Cracks were found between the intake and exhaust channels, particularly for the middle pistons of one specific model. The remanufacturer can repair smaller cracks. If the crack is too severe, the welding process may cause warping, weakening the part.

Wear is also likely to occur at the interface between cylinder heads and engine blocks. The cylinder heads may be made of a different material from the blocks and therefore have a different degree of thermal expansion, resulting in shear stress and wear along the interface. After a considerable amount of wear, the cylinder head becomes undersize and is thus scrapped.

It should be noted that the failure mode "disassembly impossible" does not occur during the disassembly of cylinder heads from engine blocks, but instead during the disassembly of valve-pin guides. Since the guides are pressed fitted, disassembly without damage is difficult. If the guides are pressed out using excessive force, damage occurs to the cylinder head, resulting in a scrap mode of "weakens part."

Crankshaft - The rotation of crankshafts relative to connecting rods causes wear at the shafts, which become undersize. Furthermore, if an engine is not regularly maintained, it may lack lubricant and become burnt. The majority of scrapped crankshafts have a failure mode of "burnt." Since a large amount of material is worn from burnt parts, these crankshafts become undersize.

On the other hand, many crankshafts were scrapped with the failure mode of "no failure" because the remanufacturer studied overstocked a particular model of crankshafts.

Camshaft - Approximately half of the total camshafts scrapped had failure mode "wear". Since pieces of material are flaked off from the cams, the scrap mode is "material loss" instead of "undersize."

Another major failure mode of scrapped camshafts is corrosion, which is more severe during the summer due to higher temperatures and humidity. The camshafts may be stored at the remanufacturer for an extended period of time before being processed.

Connecting Rod - As previously stated, in an engine lacking maintenance, parts including crankshafts and connecting rods would be burnt. Wear at the bore of the connecting rod thus becomes more severe, resulting in oversize of the bore. For connecting rods, "no failure" usually meant a mismatched cap, resulting in a scrap mode of "mating part lost". Caps usually become mismatched during disassembly. Damage to connecting rods may be made during disassembly when connecting rods are wedged inside engine blocks. Disassemblers would strike or pull connecting rods out of blocks using excessive force. To save the more valuable blocks, the less valuable connecting rods would be sacrificed.

The failure mode "disassembly impossible" may result from two possibilities. First, the connecting rods are retained in the engine blocks and cannot be removed using any available method. Second and more commonly, the connecting rods cannot be disassembled from the pistons. In both cases, the impossibility of disassembly could be due to lack of lubricant and/or extended storage time where corrosion within the blocks may cause flaked-off material to impede removal of the connecting rods. In the first case, the connecting rods have scrap mode "unknown damage" since the specific damage could not be identified. In the second case, the connecting rods have scrap mode "no process".

Oil Pan/Valve Cover - Most oil pans/valve covers have failure modes of dent and corrosion since these parts are located at the outer layer of the engine, protecting the cylinder heads and crankshafts. The pans/covers are made of sheet metal that can be easily deformed. Dents may also result during the transportation of engines to the remanufacturer since engines may be dropped on floors or conveyors. Also, oil pans/valve covers have direct contact with the surrounding environment and are vulnerable to corrosion.

Manual reshaping can be used to repair most of the dented oil pans/valve covers. However, some dents cannot be accessed for repair and result in scrap mode "no process". Corroded oil pans/valve covers have scrap mode "material loss".

Cylinder Sleeve - Cylinder sleeves guide the movement of pistons and are used in only certain types of engines. As expected, wear is the dominant failure mode.

STATISTICAL ANALYSIS

The sample of parts was statistically analyzed using a normal approximation to assess the 95 percent confidence level of the probability of assigned failure modes and scrap modes. Parts were analyzed as to their probability of incurring a single failure mode (failed/not failed) and single scrap mode (scrap/not scrap) on a part count basis.

DERIVATION OF FMEA INDICES

OCCURRENCE

Table 1 shows the percentage of each part type entering the waste stream categorized into the different failure modes. The bottom row of Table 1 represents the probability that a part entering the waste stream has a particular failure mode and is relevant to the FMEA index of occurrence (OCC). These probabilities are used to derive the OCC index as follows. The largest probability, in our case, 21.8% for "wear", is assigned an OCC value of 10, while the smallest probability, 0.1% for "handling damage" is assigned an OCC value of 1. All other probabilities are normalized to these two extremes to obtain corresponding values for OCC. Derivation of OCC values for each failure mode is shown in Table 5a.

Table 2 shows the proportion of scrap modes for each part type. These percentages are not used to derive an FMEA index, but the scrap modes do correspond to the FMEA category of effect of failure. In our case, the scrap mode corresponds to the actual, rather than the expected, effect of failure modes.

DETECTABILITY

The next FMEA index of interest is DET or detectability. In our study, we relate detectability to the point in the remanufacture process that a failure is detected. For example, it is far better that a crack is detected at disassembly, so that an immediate decision can be made whether the crack is repairable and if so, by which processes, than if that same crack were not detected until final assembly. Therefore, the later the detection, the higher the value of the DET index. Table 3 shows the proportion of all parts combined that were detected at a particular processing station with a particular failure mode.

Parts are first analyzed with respect to the processing sequence. Generally, all parts go through the stages of disassembly, cleaning, painting and inspection. After inspection, the sequence of processes depends on the part type. Processes are grouped according to the location of the process in the sequence. For example, grinding for crankshafts is grouped together with polishing for camshafts under "Process 1" since these are the first processes immediately following inspection for these part types.

Combining failure modes, 35% and 56% of all failures were detected at the disassembly and inspection stations, respectively. Except for the failure modes "design flaw", "disassembly damage", "disassembly impossible", "fracture," and "handling damage", the majority, greater than 50%, of parts for all other failure modes were detected at the inspection station.

The contrast in detectability between the stations for disassembly and inspection is substantial for wear. Specifically, over 80% of wear was detected at the inspection station, which is the highest percentage of detection at a single station for a failure mode, excluding extreme and special cases. This high contrast is because wear in small amounts is difficult to detect by the disassemblers who do not have the appropriate measurement equipment. Only severe cases of "wear", as well as "burnt", "crack", "fracture" and "hole," are detectable at the disassembly stage. Hence, the detectability of a failure mode at the inspection station depends on gauging accuracy, and the detectability of a failure mode at the disassembly station depends on the severity of the failure.

Next addressed is the derivation of the FMEA DET detectability index from the data collected. The index reflects the concept that the more parts for a particular failure mode that can be detected at an early stage, the better the detectability of that failure mode.

Each processing stage is assigned with a weight, based on the location in the sequence. Since there are a total of 9 stages defined, the earliest stage, disassembly, is assigned with a weight of 1/9 while the next stage, cleaning, is assigned with a weight of 2/9. Inspection is assigned a weight of 4/9 and the steps grouped together as "Process 1", or the step immediately following inspection for different parts, are all assigned the identical weight of 5/9. Subsequent steps grouped together as Processes 2, 3 and 4 are also identically weighted by group. The weighting factors and percentages of parts detected at each stage are first multiplied together. Then, the weighted percentages of different stages are combined to obtain the overall weighted percentage. Finally, the overall weighted percentage is multiplied by 10 to obtain a DET index with a range up to 10. Derivation of DET values for each failure mode is shown in Table 5b.

Following this method of calculating detectability, "machining damage" has the highest, least desirable index value. This failure mode is incurred during machining, a fairly late stage, and can therefore only be detected at later stages. The lateness of both the occurrence and detection increases the index, accounting for the fact that resources required to perform earlier activities on a discarded part are wasted.

Furthermore, the DET index of the failure modes "design flaw" and "handling damage" have the most favorable values. The design flaws studied were highly detectable because they can be identified from model labels on the parts. However, the detectability of handling damage calculated is not reliable because only one part, an engine block missing caps, belongs to this category.

All other failure modes have index values in the range which corresponds to moderate to highly desirable detectability. This result is reasonable because most of the failure modes are detected at the inspection station where a reasonable degree of detectability is expected due to the quality of the equipment and the skill of the inspectors.

Table 1. Failure Modes for Each Part Type

Part Type	Bent	Burnt	Corrosion	Crack	Dent	Design Flaw	Disassembly Damage	Disassembly Impossible	Fastener Failure	Fracture	Handling Damage	Hole	Machining Damage	Wear	No Failure	Part Type
Block		0.7%	4.2%	10.5%	0.7%	2.8%	1.4%	22.4%	1.4%	7.0%	0.7%	9.1%	3.5%	24.4%	11.2%	Block
95% CI		1.4%	3.3%	5.0%	1.4%	2.7%	1.9%	6.8%	1.9%	4.2%	1.4%	4.7%	3.0%	7.0%	5.2%	
Cylinder Head	2.8%	6.7%	5.6%	23.0%			7.9%	19.6%	3.4%	0.6%		1.1%	6.2%	22.5%	0.6%	Cylinder Head
95% CI	2.4%	3.7%	3.4%	6.2%			4.0%	5.8%	2.7%			1.5%	3.5%	6.1%	1.1%	
Crankshaft	0.6%	37.2%	5.0%				5.0%	3.7%	2.5%	3.1%			2.5%	20.5%	19.9%	Crankshaft
95% CI	1.2%	7.5%	3.4%				3.4%	2.9%	2.4%	2.7%			2.4%	6.2%	6.2%	
Camshaft		0.6%	13.3%	2.9%			2.3%	2.3%	4.0%	5.2%			6.4%	57.8%	5.2%	Camshaft
95% CI		1.1%	5.1%	2.5%			2.2%	2.2%	2.9%	3.3%			3.6%	7.4%	3.3%	
Connecting Rod	10.3%	26.1%	1.1%				16.9%	20.1%	2.0%	0.9%				6.3%	16.3%	Connecting Rod
95% CI	3.2%	4.6%	1.1%				3.9%	4.2%	1.5%	1.0%				2.5%	3.9%	
Oil Pan / Valve Cover			14.5%		69.8%			1.3%		5.0%		3.8%		4.4%	1.2%	Oil Pan / Valve Cover
95% CI			5.5%		7.1%			1.7%		3.4%		3.0%		3.2%	1.7%	
Cylinder Sleeve			18.3%	11.0%	1.2%			4.9%		7.3%			15.8%	41.5%		Cylinder Sleeve
95% CI			8.4%	6.8%	2.4%			4.7%		5.6%			7.9%	10.7%		
Overall percentage of parts for each failure mode	3.4%	14.5%	5.9%	5.6%	9.1%	0.3%	7.0%	12.3%	2.7%	2.7%	0.1%	1.7%	3.5%	21.8%	9.4%	Overall percentage of parts for each failure mode
95% CI	1.0%	2.0%	1.3%	1.3%	1.6%	0.3%	1.4%	1.8%	0.9%	0.9%	0.2%	0.7%	1.0%	2.3%	1.6%	95% CI

Table 2. Scrap Modes for Each Part Type

Part Type	Cosmetic	Last Oversize	Last Undersize	Makes Oversize	Makes Undersize	Material Loss	Mating Part Lost	No Process	Overstock	Sacrificial Part	Unknown Damage	Warped Part	Weakens Part	Part Type
Block		25.2%	1.4%	1.4%	0.7%	19.6%	1.4%	32.8%			16.1%		1.4%	Block
95% CI		7.1%	1.9%	1.9%	1.4%	6.5%	1.9%	7.7%			6.0%		1.9%	95% CI
Cylinder Head		7.3%	9.5%		6.2%	18.0%	0.6%	9.5%			11.8%	3.4%	33.7%	Cylinder Head
95% CI		3.8%	4.3%		3.5%	5.6%	1.1%	4.3%			4.7%	2.7%	6.9%	95% CI
Crankshaft			55.9%		2.5%	15.5%		3.1%	19.9%		3.1%			Crankshaft
95% CI			7.7%		2.4%	5.6%		2.7%	6.2%		2.7%			95% CI
Camshaft			20.8%		11.0%	61.3%	0.6%	2.9%			1.7%		1.7%	Camshaft
95% CI			6.0%		4.7%	7.3%	1.1%	2.5%			1.9%		1.9%	95% CI
Connecting Rod		31.8%			3.2%	18.1%		8.3%		16.9%	12.0%		9.7%	Connecting Rod
95% CI		4.9%			1.8%	4.0%		2.9%		3.9%	3.4%		3.1%	95% CI
Oil Pan / Valve Cover	0.6%					18.9%	3.8%	76.7%						Oil Pan / Valve Cover
95% CI	1.2%					6.1%	3.0%	6.6%						95% CI
Cylinder Sleeve	1.2%	28.0%		15.9%		24.4%		23.2%			4.9%		2.4%	Cylinder Sleeve
95% CI		9.7%		7.9%		9.3%		9.1%			4.7%		3.3%	95% CI
Overall percentage of parts for each scrap mode	0.2%	14.7%	11.6%	1.2%	2.8%	20.2%	5.9%	19.6%	2.6%	4.7%	7.9%	0.5%	8.1%	Overall percentage of parts for each scrap mode
95% CI	0.2%	2.0%	1.8%	0.6%	0.9%	2.2%	1.3%	2.2%	0.9%	1.2%	1.5%	0.4%	1.5%	95% CI

Table 3. Processes where Failure Modes Detected

Failure Mode	1 Disassembly	2 Clean	3 Paint	4 Inspection	5 Block: Threading	5 Cylinder Head: Remove Guide	5 Crankshaft: Grinding	5 Camshaft: Polishing	5 Connecting Rod: Remove Fastener	5 Oil Pan/Valve Cover: Reshaping	5 Cylinder Sleeve: Boring	6 Block: Seat Cutting	6 Cylinder Head: Threading	6 Crankshaft: Polishing	6 Connecting Rod: Polishing	6 Oil Pan/Valve Cover: Painting	7 Block: Boring	7 Cylinder Head: Boring	7 Connecting Rod: Weighting	8 Block: Final Inspection	8 Cylinder Head: Milling	9 Assembly	Failure Mode
Bent	40.5%			59.5%																			Bent
95%CI	14.8%			14.8%																			95%CI
Burnt	21.7%			78.3%																			Burnt
95%CI	6.0%			6.0%																			95%CI
Corrosion	24.3%			70.3%				4.1%													1.3%		Corrosion
95%CI	9.8%			10.4%				4.5%													2.6%		95%CI
Crack	34.3%			58.6%													5.7%	1.4%					Crack
95%CI	11.1%			11.5%													5.4%	2.8%					95%CI
Dent	22.1%			77.9%																			Dent
95%CI	7.7%			7.7%																			95%CI
Design Flaw	100.0%																						Design Flaw
95%CI	0.0%																						95%CI
Disassembly Damage	88.5%			9.2%	2.3%																		Disassembly Damage
95%CI	6.7%			6.1%	3.1%																		95%CI
Disassembly Impossible	87.6%			5.9%	6.5%																		Disassembly Impossible
95%CI	5.2%			3.7%	3.9%																		95%CI
Fastener Failure	3.0%			64.7%					26.5%				2.9%						2.9%				Fastener Failure
95%CI	5.7%			16.1%					14.8%				5.7%						5.7%				95%CI
Fracture	70.6%			29.4%																			Fracture
95%CI	15.3%			15.3%																			95%CI
Handling Damage	100.0%																						Handling Damage
95%CI	0.0%																						95%CI
Hole	47.6%			52.4%																			Hole
95%CI	21.4%			21.4%																			95%CI
Machining Damage							9.1%	36.4%			15.9%	4.5%					9.1%	11.4%			13.6%		Machining Damage
95%CI							8.5%	14.2%			10.8%	6.2%					8.5%	9.4%			10.1%		95%CI
Wear	4.4%			81.2%								0.4%					12.9%	0.4%			0.7%		Wear
95%CI	2.4%			4.7%								0.7%					4.0%	0.7%			1.0%		95%CI
No Failure	41.0%			58.1%																	0.9%		No Failure
95%CI	8.9%			8.9%																	1.7%		95%CI
Overall Percentage of parts Detected at Each Process	34.9%			55.8%	1.0%	0.3%		1.5%	0.7%		0.6%	0.2%	0.1%				3.5%	0.7%			0.7%		Overall Percentage of parts Detected at Each Process
95%CI	2.6%			2.8%	0.5%	0.3%		0.7%	0.5%		0.4%	0.3%	0.2%				1.0%	0.5%			0.5%		95%CI

16

Table 4. Repairablity of Failure Modes

Part Type		Bent	Burnt	Corrosion	Crack	Dent	Design Flaw	Disassembly Damage	Disassembly Impossible	Fastener Failure	Fracture	Handling Damage	Hole	Machining Damage	Wear	No Failure	Part Type
Block	Repairable																Block
	95%CI																
	Unrepairable																
	95%Cf																
Cylinder Head	Repairable			43.1%	78.1%				24.5%	26.3%	50.0%				16.1%		Cylinder Head
	95%CI			12.7%	5.7%				12.0%	19.8%	49.0%				9.6%		
	Unrepairable	100.0%	100.0%	17.2%	21.9%			100.0%	71.4%	31.6%	50.0%		25.0%	100.0%	71.4%	33.3%	
	95%CI	0.0%	0.0%	9.7%	5.7%			0.0%	12.6%	20.9%	49.0%		34.6%	0.0%	11.8%	53.3%	
Crankshaft	Repairable																Crankshaft
	95%CI																
	Unrepairable																
	95%CI																
Camshaft	Repairable																Camshaft
	95%CI																
	Unrepairable																
	95%CI																
Connecting Rod	Repairable																Connecting Rod
	95%CI																
	Unrepairable																
	95%CI																
Oil Pan / Valve Cover	Repairable					93.6%											Oil Pan / Valve Cover
	95%CI					1.2%											
	Unrepairable			39.7%		6.4%				4.1%	42.1%		75.0%		12.5%	66.7%	
	95%CI			12.6%		1.2%				5.5%	22.2%		34.6%		8.7%	53.3%	
Cylinder Sleeve	Repairable																Cylinder Sleeve
	95%CI																
	Unrepairable																
	95%CI																
Overall number of parts for each failure mode	Repairable			43.1%	78.1%	93.6%			24.5%	26.3%	50.0%				16.1%		Overall number of parts for each failure mode
	95%CI			12.7%	5.7%	1.2%			12.0%	19.8%	49.0%				9.6%		
	Unrepairable	100.0%	100.0%	56.9%	21.9%	6.4%		100.0%	75.5%	73.7%	50.0%	100.0%	100.0%	100.0%	83.9%	100.0%	
	95%CI	0.0%	0.0%	12.7%	5.7%	1.2%		0.0%	12.0%	19.8%	49.0%	0.0%	0.0%		9.6%	0.0%	

Tables 5a,b,c. Derivation of OCC, DET and REP Indices

Failure Mode	% Waste Stream	OCC
Bent	3.4%	2.4
Burnt	14.5%	7.0
Corrosion	5.9%	3.4
Crack	5.6%	3.3
Dent	9.1%	4.7
Design Flaw	0.3%	1.1
Disassembly Damage	7.0%	3.9
Disassembly Impossible	12.3%	6.1
Fastener Failure	2.7%	2.1
Fracture	2.7%	2.1
Handling Damage	0.1%	1.0
Hole	1.7%	1.7
Machining Damage	3.5%	2.4
Wear	21.8%	10.0
No Failure	9.4%	4.9

Failure Mode	Weighted %	DET
Bent	31.0%	3.1
Burnt	37.2%	3.7
Corrosion	37.3%	3.7
Crack	35.4%	3.5
Dent	37.1%	3.7
Design Flaw	11.1%	1.1
Disassembly Damage	15.2%	1.5
Disassembly Impossible	16.0%	1.6
Fastener Failure	48.0%	4.8
Fracture	20.9%	2.1
Handling Damage	11.1%	1.1
Hole	28.6%	2.9
Machining Damage	65.2%	6.5
Wear	47.8%	4.8
No Failure	31.1%	3.1

Failure Mode	% Not Repaired	REP
Bent	100.0%	10.0
Burnt	100.0%	10.0
Corrosion	56.9%	5.7
Crack	21.9%	2.2
Dent	6.4%	0.6
Design Flaw	100.0%	10.0
Disassembly Damage	100.0%	10.0
Disassembly Impossible	75.5%	7.6
Fastener Failure	73.7%	7.4
Fracture	50.0%	5.0
Handling Damage	100.0%	10.0
Hole	100.0%	10.0
Machining Damage	100.0%	10.0
Wear	83.9%	8.4
No Failure	100.0%	10.0

Table 6. Derivation of Risk Priority Number (RPN)

Failure Mode	OCC	DET	REP	RPN
Bent	2.4	3.1	10.0	73.4
Burnt	7.0	3.7	10.0	258.0
Corrosion	3.4	3.7	5.7	71.8
Crack	3.3	3.5	2.2	25.3
Dent	4.7	3.7	0.6	10.5
Design Flaw	1.1	1.1	10.0	12.1
Disassembly Damage	3.9	1.5	10.0	57.9
Disassembly Impossible	6.1	1.6	7.6	73.7
Fastener Failure	2.1	4.8	7.4	73.8
Fracture	2.1	2.1	5.0	21.8
Handling Damage	1.0	1.1	10.0	11.0
Hole	1.7	2.9	10.0	48.2
Machining Damage	2.4	6.5	10.0	156.7
Wear	10.0	4.8	8.4	403.2
No Failure	4.9	3.1	10.0	150.6

REPAIRABILITY

In the FMEA adapted for remanufacture, the severity index (SEV) has been renamed repairability (REP) to reflect the severity of a failure mode for a remanufacturer. Repairability reflects the degree to which parts are successfully repaired by the remanufacturer. Consistent with the index values for SEV, the higher the index for REP the lower the repairability. Table 4 shows the proportion of parts that are repaired, as well as the 95% confidence interval of each entry. The percentage of parts that is not repaired is used to derive the REP indices for failure modes as shown in Table 5c.

Since the goal of the OER is to assemble a number of remanufactured engines per day, week or month, if a part requires a longer-than-usual time to repair, it would be scrapped for cost effectiveness. This explains why, in general, OER has a low repair rate of damaged parts. Only two types of engine parts are truly repaired at the OER - cylinder heads with mainly cracks and/or corrosion and oil pans/valve covers with dents.

Cracks on cylinder heads are usually located between the valve seats due to the high stress in this region. To repair cracks, some material surrounding the cracks is first removed. These regions are then welded and machined back to shape. Voids resulting from corrosion are processed using the same procedure. Dents on oil pans/valve covers are removed unless the back of dent is inaccessible.

Even though the OER repairs only these two types of parts, it is possible that a number of other failures may be repaired as follows.

Bent - If a connecting rod is not severely bent, it can be mechanically straightened. After straightening, reliability testing is required.

Corrosion - Corrosion on crankshafts may be repaired by spot welding. Corrosion may also be repaired using liquid-metal filler on engine blocks that are otherwise scrapped due to cosmetic reasons.

Crack - Cracks between water jackets on engine blocks may be repaired by fitting inserts to block leakage through the crack.

Fastener Failure - Worn-out threading may be repaired using inserts. First, the threaded hole is drilled to a larger size. Next, an insert is installed and the desired threading is provided by the insert.

Fracture - Fractured towers on cylinder heads may be repaired using welding.

Wear – Worn, oversized piston housings in engine blocks may be repaired by inserting and machining a sleeve to specifications. Scratches on the journal of crankshafts may be repaired using welding. However, if there are scratches on more than one journal, the repair cost may exceed the value of the crankshaft.

At the OER, failure modes that result in removal of material are scrapped since repair would not be possible without a material-addition repair process. In many cases, welding is not used to repair parts that experience high stresses during engine operation. For instance, burns cannot be repaired since excessive material is usually worn from burnt parts. Corrosion and scratches on camshafts are not repaired using welding because camshafts are subjected to high compressive stresses during engine operation and the durability of the welding is not reliable. Worn journals on crankshafts, camshafts and connecting rods are not repaired since replacement of material is necessary. Due to their low economic value, connecting rods are typically not repaired. Machining damage is not likely to be repaired since no process is used that can replace the amount of material that is mistakenly removed.

RISK PRIORITY NUMBER

Finally, the risk priority numbers (RPN) for each failure mode are determined by finding the product of OCC, DET and REP, as shown in Table 6. Table 6 shows that the RPN for "wear" and "burnt" are highest since these failure modes are very common and cannot be repaired without a material-replacement process. Furthermore, detection of material removal beyond specifications is typically not detected until the inspection stage of the remanufacturing process. Therefore, "wear" and "burnt" are the failure modes that cause the most difficulty for remanufacturers. Product or process design that decreases the occurrence, detection or repairability indices for "wear" and "burnt" would facilitate remanufacture by decreasing the portion of the waste stream that have these failure modes.

SUMMARY

The goal of this work is to enable the more efficient remanufacture of products through design. Since remanufacturers aim to reuse parts, parts that enter their waste stream reveal difficulties in remanufacture. The waste stream of an original-equipment automotive remanufacturer was analyzed to gain insight into reasons why parts are not reused. The data gathered were used to derive values for the indices of occurrence (OCC), detectability (DET) and repairability (REP) for an FMEA modified for remanufacture. Product or process design that aims to reduce the Risk Priority Number (RPN) of the failure modes identified would facilitate remanufacture.

ACKNOWLEDGMENTS

The authors gratefully acknowledge the financial support of the Natural Sciences and Engineering Research Council of Canada (NSERC).

REFERENCES

1. Eubanks, C.F., Kmenta, S., Ishii, K., 1996, "System Behavior Modeling as a Basis for Advanced Failure Modes and Effects Analysis", Proc. ASME Design Eng. Tech. Conf. and Comp. in Eng. Conf., Irvine, CA, U.S.A, DETC96/CIE1340.
2. Finger, S., Dixon, J.R., 1989, "A Review of Research in Mechanical Engineering Design. Part 2: Representations, Analysis, and Design for Life Cycle", Research in Engineering Design, 1:121-137.
3. Hammond, R., Amezquita, T., and Bras, B.A., 1998, "Issues in the Automotive Parts Remanufacturing Industry: Discussion of Results from Surveys Performed among Remanufacturers," Int. J. of Engineering Design and Automation, 4/1:27-46.
4. Hammond, R., and Bras, B.A., 1996, "Design for Remanufacturing Metrics," Proc. 1st Int. Workshop on Reuse, Eindhoven, The Netherlands, pp. 5-22.
5. Ishii, K., Lee, B.H., 1996, "Reverse Fishbone Diagram: A Tool in Aid of Design for Product Retirement", Proc. ASME Design Eng. Tech. Conf. and Comp. in Eng. Conf., Irvine, CA, U.S.A, DETC96/DFM-1272.
6. Kimura F., Hata, T., Suzuki H., 1998, Product Quality Evaluation Based on Behaviour Simulation of Used Products, Annals of the CIRP, 47/1:119-122.
7. Kmenta S., Fitch, P., Ishii K., 1999, "Advanced Failure Modes and Effects Analysis of Complex Processes," Proc. ASME Design Eng. Tech. Conf. and Comp. in Eng. Conf., Las Vegas, NV, U.S.A., DETC99/DFM8939.
8. Lund, R., 1996, The Remanufacturing Industry: Hidden Giant, Boston University, Boston, MA, U.S.A.
9. SAE, 1995, "Potential Failure Mode and Effects Analysis (FMEA)," SAE J-1739.
10. Mangun, D., Thurston, D.L., 2000, "Product Portfolio Design for Component Reuse," Proc. IEEE Int. Symp. Electronics and the Environment, San Francisco, CA, USA, pp. 86-92.
11. Sherwood, M., Shu, L., "Modified FMEA using Analysis of Automotive Remanufacturer Waste Streams to Support Design for Remanufacture," in the Proceedings of the 2000 ASME DETC and CIE Design Theory and Methodology Conference, Baltimore, MD, USA, September 10-14, 2000, DETC2000/DTM-14567.
12. Shu, L., Flowers, W., 1998, Reliability Modeling in Design for Remanufacture, Trans. of the ASME Journal of Mechanical Design, 120/4:620-627.
13. Shu, L., Flowers, W., 1999, Application of a Design-for-Remanufacture Framework to the Selection of Product Life-Cycle Fastening and Joining Methods, Int. J. Robotics and Computer Integrated Manufacturing, 15/3:179-190.
14. Stamatis, D., 1995, Failure Mode and Effects Analysis: FMEA from Theory to Execution. Milwaukee, WI: ASQC Quality Press.
15. U.S. Congress, 1992, Green Products by Design: Choices for a Cleaner Environment, OTA-E-541, Office of Technology Assessment, Washington, D.C.
16. Umeda, Y., Shimomura, Y., Yoshioka, M., Tomiyama, T., 1999, "A Proposal of Design Methodology for Upgradable Products," Proc. ASME Design Eng. Tech. Conf. and Comp. in Eng. Conf., Las Vegas, NV, U.S.A., DETC99/DFM8969.
17. Williams, J., Shu, L., 2000, "Analysis of Toner-Cartridge Remanufacture Waste Stream," Proc. IEEE Int. Symp. Electronics and the Environment, San Francisco, CA, USA, pp. 260-265.
18. Zussman, E., Kriwet, A., and Seliger, G., 1994, "Evaluation of Product Ease of Separation into Materials and Components in the Recycling Process," Annals of the CIRP, 43,1:9-1.

CONTACT

This paper described the Master of Engineering project research of A. Lam and the Master of Applied Science thesis research of M. Sherwood at the Life-Cycle Design Laboratory, Department of Mechanical and Industrial Engineering, University of Toronto.

This work was supervised by L. Shu, whose Ph.D. from the Department of Mechanical Engineering at MIT is in the area of environmentally conscious product design, specifically design for remanufacture.

Corresponding author:
Prof. L. Shu
Department of Mechanical and Industrial Engineering
University of Toronto
5 King's College Road
Toronto, Ontario, M5S 3G8 Canada
E-mail: shu@mie.utoronto.ca
Tel: 416 946 3028

2001-01-0341

Environment-Friendly Fluxless Soldering Process for High Sealing Ability on Pressure Sensors

Takao Yoneyama, Hiroyuki Okada, Katsuhiro Izuchi and Fumio Kojima
Denso Corp.

ABSTRACT

In a conventional soldering process, solvents, such as chlorofluorocarbons (CFCs), have been necessary to remove the flux-residue after soldering.
A new CFC-free fluxless soldering process has been developed to obtain high sealing ability even in a small soldering area. This new process utilizes a reducing atmosphere with an appropriate load and assembly orientation to solder the parts. Under this fluxless condition, it is found that appropriate loading and good solder-wettability of the upper part increase the wettability of the lower part.

INTRODUCTION

It is important to keep high sealing ability with a small soldering area, especially for pressure sensors. Additionally, it is necessary to maintain good solder-wettability for retaining high sealing ability.
In a conventional soldering process, soldering flux has been used to obtain good wettability. Solvents, such as CFCs, have been necessary to remove the flux-residue after soldering in order to obtain the essential sensing property. CFCs, however, pose an environmental concern and must be eliminated to protect the ozone layer.
To solve this problem, a new fluxless soldering process has been developed to obtain high sealing ability even for small soldering areas. This new soldering technique is not only CFC-free, but it is also easy to incorporate within an automatic assembly process.
Solder-wettability phenomena without flux is currently not well understood. Therefore, this paper describes the procedure which has been found to improve the solder-wettability in a fluxless soldering process.

FACTORS CONTROLLING SOLDER-WETTABILITY

ROLE OF FLUX AND ITS SUBSTITUTION – The conventional soldering mechanism is generally explained as shown in Fig.1. The solder disperses or spreads so as to balance the interfacial tensions among the liquid solder, the base metal, and the vapor phase[1].

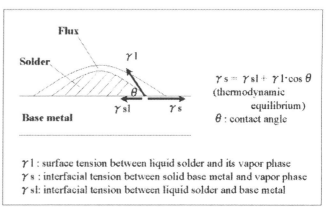

$$\gamma s = \gamma sl + \gamma l \cdot \cos \theta$$
(thermodynamic equilibrium)
θ : contact angle

γl : surface tension between liquid solder and its vapor phase
γs : interfacial tension between solid base metal and vapor phase
γsl: interfacial tension between liquid solder and base metal

Fig.1 Soldering mechanism

The role of flux is to (1) reduce and remove oxide (from the base metal surface), (2) prevent re-oxidation during soldering, and (3) decrease the surface tension of the solder (that is, increase the spreading force).
To utilize a fluxless condition, another method must be devised to substitute for the properties of the flux as stated above. A typical reducing atmosphere of hydrogen and nitrogen (H_2+N_2) is employed.
NEW FLUXLESS SOLDERING APPLICATION – Firstly, to examine the possibility of fluxless soldering, initial experiments were conducted with the following conditions.
The typical structure of a pressure sensor is shown in Fig.2. The soldering area is 0.8mm long. A vacuum is kept inside of the cap as a sensing standard.
Fluxless soldering testing was done in the H_2+N_2 atmosphere with the same parts as used in flux soldering. The soldering method was as follows: solder foil was set between upper and lower parts, and then heated in the H_2+N_2 atmosphere furnace (Fig.3).

H_2 content and temperature were varied from 30 to 50 %, and from 270 to 350 °C respectively, but good solder-wettability was not obtained.

On the other hand, when the setting orientation of the parts was changed, solder-wettability was found to improve.

Therefore, the solder spreading phenomenon was investigated more closely using a sandwich structure under fluxless conditions.

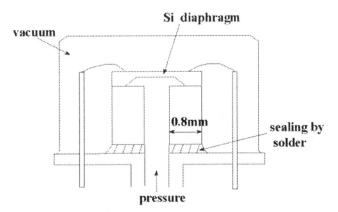

Fig.2 Typical structure of pressure sensor (cross section)

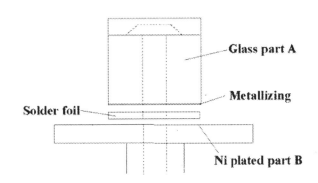

Fig.3 Setting direction of parts in the initial experiment

EXPERIMENTAL DETAILS

The sandwich structure used was as follows: gold (Au) coated glass part A (upper part), solder foil, nickel (Ni) plated part B (lower part).

EFFECT OF LOAD – Au was evaporated to a thickness of 80 nm on glass part A with a soldering area of 40 mm^2. The sandwich structure was then heated in a H_2+N_2 atmosphere (30% H_2) to a temperature of 300 °C at different levels of load. After soldering, the sandwich structure was separated at the soldered position, and the solder spreading area was measured off of a microscopic photograph. The solder contact angle θ was measured in the same manner using a cross section of another sample.

For comparison, a Au-plated part B' were also examined instead of Ni-plated part B.

EFFECT OF UPPER PART'S SOLDER-WETTABILITY – Increasing the Au thickness of the glass part A was found to increase the solder-wettability of the glass part A. The load was kept constant at 6 dyne/mm^2. Ni-plated part B was used to complete the sandwich structure which was then evaluated using the same basic experimental procedures as used to test the effect of load.

EFFECT OF PART ORIENTATION - By changing the setting orientation of the parts, the applied load and upper part's solder-wettability are believed to be changed.

The setting orientation for case 1 is A* / solder foil / B*; whereas, the setting orientation for case 2 is B* / solder foil / A*. The glass part A* (weight of 0.055g) has a soldering area of 9 mm^2 with a Au layer thickness of 100nm. The Ni-plated part B* (weight 2.75g) has better solder-wetterbility than glass part A*. After soldering, the solder wetting area was examined as described above and the sealing ability which was measured using a krypton(Kr[85]) radioisotopic method.

RESULTS

EFFECT OF LOAD – The solder spreading area on glass part A is shown in Fig.4. As anticipated, an increase in the applied load resulted in an increase in the solder spreading area.

Furthermore, the solder contact angle θ (which is determined by the surface condition) was almost constant (about 50 degrees) for this set of experiments. This indicates that load has an effect on solder spreading even if the surface condition is the same.

The sandwich structure utilizing the Au-plated part B', yielded better solder-wettability than the corresponding Ni-plated part. This result suggests that, in addition to the applied load, the opposite part's wettability also affects the solder spreading area.

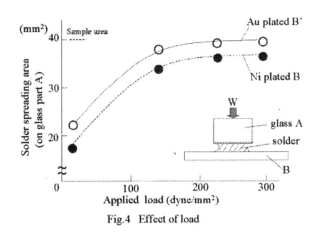

Fig.4 Effect of load

EFFECT OF UPPER PART'S SOLDER-WETTABILITY - When the Au film thickness of the glass part A (upper position) was varied, the solder spreading area on Ni-plated part B (lower position) was as shown in Fig.5. Solder-wettability of Ni-plated part B is found to increase with an increase in the wettabilty (that is, Au film thickness) of the glass part A. In other words, it is found that as the upper part's wettability increases, the lower part's wettability is also increased.

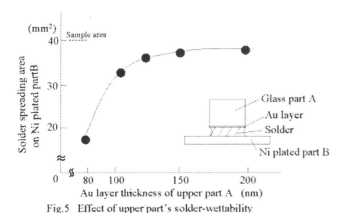

Fig.5 Effect of upper part's solder-wettability

EFFECT OF PART ORIENTATION - The solder wetting area after separation at the soldered position is shown in Fig.6. (Glass part A* is square with a hole in the center. Bright areas are voids, indicating areas that were not wetted by the solder.)
Case 2 resulted in better solder-wettability than in case 1. That is, better solder-wettability can be obtained when the upper part has better solder-wettability as compared to the lower part and a larger load can be used.

High sealing ability was accomplished in case 2. The leak rate of case 2 was below 1×10^{-6} Pa · cm^3/s, a value that is comparable to the conventional soldering process.

DISCUSSION

SOLDER WETTABILITY - From the aforementioned results, the factors which control solder-wettability are thought to be as detailed in Fig.7. In a fluxless process, it was found that both the load and the setting orientation of the parts are important. This is in addition to traditionally known factors such as soldering atmosphere, surface cleanliness, and heating method and condition.
The mechanisms for increasing solder-wettability in a fluxless process for a sandwich structure are thought to be as shown in Fig.8.
In addition to the effect of surface tensions (which is generally accepted), load and opposite surface wettability also influence the solder spreading force.
Moreover, the impact of opposite-surface wettability is thought to be caused by both the viscosity and surface tension of the solder.

In other words, although the effects of load and surface wettability balance are negligible in a conventional flux soldering process, they are thought to have a large effect in fluxless soldering.
EFFECT OF FLUXLESS SOLDERING PROCESS - By utilizing the fluxless soldering process, approximately 4500 kg/year of CFCs have been eliminated from our company's production processes (Fig.9).

	Case 1	Case2
Setting orientation	upper:Glass part (A*) →load: 6dyne/mm² lower:Ni plated part (B*)	upper:Ni plated part (B*) →load: 300dyne/mm² lower:Glass part (A*)
Glass part (A*)side	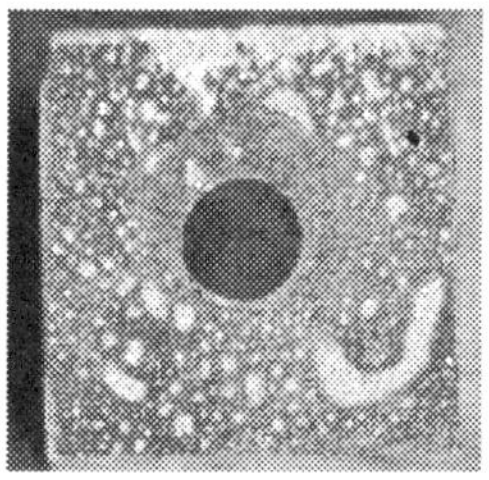	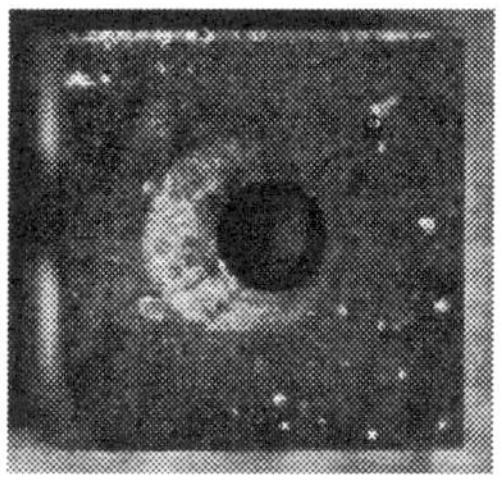
Ni plated part(B*)side		

Fig.6 Effect of part orientation

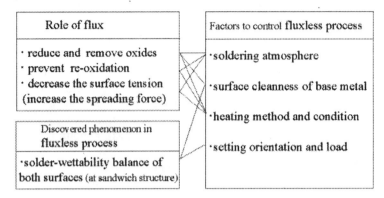

Fig.7 Factors which control solder-wettability

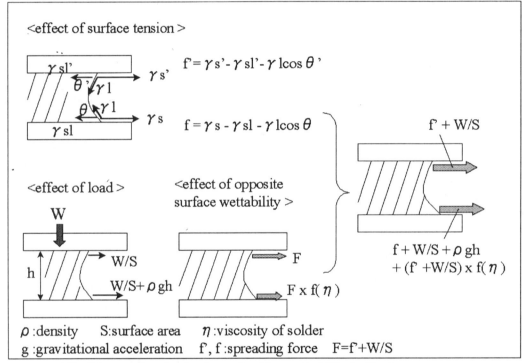

Fig.8 Increasing mechanism of solder-wettability in fluxless process

	conventional soldering process	fluxless soldering process
environmental impact	CFC - necessary	CFC - free (- 4500 kg/year)
sealing ability (gas leakage)	below 1×10^{-6} Pa·cm³/s	below 1×10^{-6} Pa·cm³/s
automatic assembly	difficult (flux coating)	easy

Fig.9 Comparison of conventional and fluxless process

CONCLUSION

1. An environmentally-friendly fluxless soldering process with high sealing ability has been developed.
2. The mechanism employed is to keep a good balance of solder-wettability between the upper and lower part's surfaces, by utilizing the appropriate load and setting orientation.

3. Under this fluxless condition, it is found that the appropriate load and good solder-wettability of the upper part will increase the wettability of the lower part [2].

REFERENCES

1. H.H.Manko, "SOLDERS AND SOLDERING", McGRAW-HILL,Inc.(1964)
2. T.Yoneyama et al., US Patent No.5289964(Mar.1994) and 5829665(Nov.1998)

Wet Versus Dry Turning: A Comparison of Machining Costs, Product Quality, and Aerosol Formation

Nathan King, Lucas Keranen, Kenneth Gunter and John Sutherland
Michigan Technological Univ.

ABSTRACT

The use of cutting fluid in machining operations not only poses a health risk to workers but also creates environmental challenges associated with fluid treatment and disposal. In an effort to minimize these concerns and eliminate the costs associated with cutting fluids, e.g., purchase, maintenance, and treatment, dry machining is increasingly being considered as an alternative. This paper is focused on comparing dry and wet machining approaches from several perspectives, including air quality, product quality, and economics. Both experimental and analytical work is presented. Experiments have been performed to determine the effect cutting fluid has on product quality and aerosol generation in the wet and dry turning of gray cast iron. To compare costs in wet and dry turning, a cost model, which includes cutting fluid-related components, has also been established. The model predictions and experimental observations are compared/contrasted to other recent reports from the technical literature.

INTRODUCTION

The use of cutting fluids in machining operations can provide a number of process benefits such as increased tool life, improved surface finish, reduced cutting forces, enhanced chip removal, and improved dimensional accuracy. However, despite the many process advantages, cutting fluids also pose serious economic and environmental problems. For instance, the use of cutting fluid requires additional capital expenditures for fluid purchase, maintenance, and disposal and can account for a significant portion of the total machining cost. Furthermore, even after treatment and disposal, cutting fluids can pollute lakes, streams, and ground water reserves, while exposure to cutting fluids, especially in the form of fluid mist, can cause adverse health effects, such as bronchitis, occupational asthma, and overall decreased lung function. In addition, workers exposed to cutting fluid mist often exhibit an increased risk to certain types of cancers, such as cancer of the stomach, pancreas, esophagus, and larynx [1,2].

Because of such health problems a number of government agencies such as the National Institute of Occupational Safety and Health (NIOSH), the Environmental Protection Agency (EPA), and the Occupational Safety and Health Administration (OSHA) have become involved in establishing standards and regulations for particulate exposure. Standards set by industrial organizations and government agencies closely follow the U.S. National Ambient Air Quality Standards (NAAQS) established by the EPA. In 1987 the NAAQS set maximum mass concentrations levels for PM10, airborne particulate matter less than 10 microns. This represents the thoracic fraction of particulate matter, which is the portion of inhalable particles that pass the larynx and penetrate into the conducting airways (e.g., trachea) and the bronchial region of the lungs. Particles larger than 10 microns are naturally expelled from the body in a relatively short amount of time and therefore present a lesser health threat. In July 1997 the EPA, in response to a growing concern that smaller particles pose a greater risk to human health, revised the NAAQS to include a PM2.5 standard [3]. Particulate matter less than 2.5 microns (PM2.5) represents the respirable fraction of inhalable particles, which are those particles that enter the non-ciliated alveoli or the deepest part of the lungs.

Current OSHA regulations require manufacturers to meet a permissible exposure level (PEL) of 5 mg/m^3 (8-hour time weighted average (TWA)) for mineral oil mist and 15 mg/m^3 for "particulates not otherwise regulated", such as cast iron dust and other machining fluid mist. However, these regulations will likely become more and more strict, considering the increasing amount of knowledge regarding the ill health effects of metalworking fluids and the external pressures of organizations such as the United Auto Workers (UAW). In fact, the UAW is currently negotiating with Ford, Chrysler, and General Motors about reducing the acceptable level of cutting fluid mist exposure to 0.5 mg/m^3. If put into affect, this is expected to cost Ford, Chrysler, and General Motors over $1.5 billion in renovation and new equipment costs [4]. Similarly, NIOSH also recommends an exposure limit of 0.5 mg/m^3 (10-hour TWA) for all metalworking fluids [5].

Because of its potential for reducing the economic and environmental burdens frequently associated with coolant use, dry machining, which involves machining in the absence of or with very small amounts of cutting fluid, is gaining favor throughout industry as an alternative to wet machining. Machining processes that are run dry virtually eliminate coolant-related costs, while at the same time eliminating the adverse worker health effects and environmental burdens associated with coolant use and disposal. However, depending on such factors as the specific machining operation, the cutting conditions, and the workpiece material, the switch from wet to dry machining may or may not be economically and/or environmentally beneficial.

There are also a number of serious concerns regarding dry machining. If cutting fluid is fundamental to the machining process itself, eliminating coolant could very well have detrimental effects on the size, finish, and shape of the finished part. Also, while dry machining does not generate mist, it may produce noticeable amounts of suspended dust, which also has the propensity to cause adverse health effects. In fact, the EPA indicates that humans exposed to even small concentrations of certain metals via an inhalation mechanism may suffer acute pulmonary effects. For example, occupational exposure to arsenic, beryllium, copper, vanadium, chromium, and zinc has been reported to produce adverse health effects ranging in severity from mild respiratory distress to severe lung dysfunction, and even cancer. The inhalation of iron dusts also has the potential to cause negative health effects and often results in a lung condition known as siderosis, which may involve the formation of numerous pulmonary nodules with limited fibrosis [6]. Therefore deciding whether or not to use cutting fluid in a particular machining process can be a complicated decision. The machining costs, waste streams, product quality, and aerosols generated may have important implications in determining whether to implement a wet or dry machining process. This paper describes efforts to integrate the economic, product quality, and environmental/health issues associated with wet and dry machining by analyzing a particular cutting operation, specifically the turning of gray cast iron.

A comparative cost model has been developed by itemizing all costs associated with machining, including those specifically related to cutting fluid purchase, maintenance, and disposal. The implementation of the cost model is illustrated by including a case study involving the turning of cast iron shafts. Also, in order to better understand the effects cutting fluid has on process performance and product quality, a set of designed experiments were conducted to study the effect a synthetic cutting fluid has on the cutting forces, surface finish, and dimensional error associated with the turning of gray cast iron. In addition, designed experiments were conducted to characterize and compare the aerosols produced in wet and dry turning. Specific attention was paid to the amount and size of the dust and mist produced, as well as to the important factors in aerosol generation. Experimental efforts were also made to better understand the mechanism behind mist formation, since such insight may be beneficial in developing new methods of mist control.

MACHINING COSTS

An economic model, which includes costs associated with power, labor, tooling, and equipment, as well as those costs associated specifically with coolant use, was developed for the purpose of comparing the unit costs associated with wet and dry machining. An effective cost comparison for the turning of

gray cast iron was performed, by itemizing and summing all costs individually associated with wet and dry cutting. It is important to note, however, that this cost model was based on unit costs, or those costs attributable to a single machining operation. This is of great significance since many finished parts require a series of successive machining operations or a machining line, rather than simply a single cutting process. Although there are some factors that are unique to a machining line, the unit costs for each machining operation used in producing a component, can, with some manipulation, be added to provide an estimate of the total machining cost.

In the research presented here, the economic model is illustrated by conducting a case study, involving the single pass turning of cast iron shafts. All the equipment and cutting fluid-related costs and parameters used in the case study were selected based on surveys conducted with cutting fluid suppliers, equipment vendors, and machining facilities. Additional cost-related information was obtained from specific machining costs reported throughout the current technical literature. Tables 1 and 2 specify some of the important costs, equipment, and cutting fluid related parameters used in the turning study.

Machine Type	Cincinnati Milacron Chucking Center
Insert Type	KC9120 Kennametal Carbide Inserts
Feed	0.15 mm/rev
Cutting Speed	427 m/min
Depth of Cut	1.52 mm
Material Type	ISO R185, Grade 20, gray cast iron
Initial Diameter	63.50 mm
Final Diameter	60.45 mm
Length of Cut	190.50 mm
Tool Life	Wet = 30 min., Dry = 25 min.

Table 1: Machine & Workpiece Parameters.

Cutting Fluid Type	Chesterton 380 Machinery Coolant
Diluent	Water
Fluid Concentration	5 %
Purchase Price of Cutting Fluid	$7.40 / liter
Treatment / Disposal Cost	$0.60 / liter
Sump Volume	380 liters
Make-up % per day	5 %
Coolant Disposed of Annually	11,360 liters / year

Table 2: Cutting Fluid Costs & Parameters.

Upon analyzing the results of the cost comparison, a number of important observations were made. For instance, the coolant-related costs in wet turning, which include all costs associated with fluid purchase, maintenance, and disposal, were found to account for as much as 20% of the total machining costs. However, tool costs, on the other hand, were found to account for only 3% of the total production cost. This relates well to the results reported in a recent German study, in which coolant-related costs were found to frequently account for up to 17% of the total machining cost [7]. Since the coolant-related costs readily account for a large portion of the total machining costs, especially in comparison with tool costs, the benefit of increased tool life, alone, is likely not enough to warrant the use of cutting fluids. Figures 1 & 2 below illustrate the cost breakdowns for both wet and dry turning.

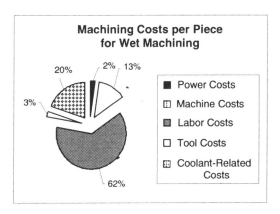

Figure 1: Production Costs in Wet Machining.

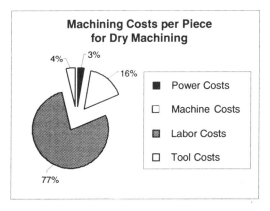

Figure 2: Production Costs in Dry Machining.

For the case study considered, it was assumed a machine operator must be present during the entire cutting process. This is why labor costs account for such a large portion of the total machining costs. This assumption may be true for small machine shops, but not for large machining facilities. Large companies typically incorporate a great deal of automated equipment and therefore do not require nearly as much skilled labor. Therefore, the cost analysis for large-scale facilities may look considerably different, with significantly decreased labor costs, but with increased equipment costs.

The ultimate goal of the cost model was to compare the unit costs in wet and dry machining, since such information is vital in determining whether wet or dry machining is the more economically sound alternative. As shown in Figure 3, the unit cost in dry turning was found to be noticeably less than that in wet turning for the conditions considered. Therefore, in this study, the additional capital expenditures required for fluid purchase, handling, and disposal far outweighed any decrease in tool cost brought about by the application of cutting fluid. Although not considered in this case study, additional equipment such as pneumatic air lines may be required to successfully facilitate the dry machining process. The capital expenditures for such dry machining equipment may have important economic ramifications and can discourage a switch from wet to dry machining.

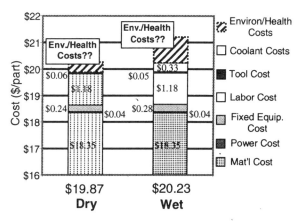

Figure 3: Unit Costs in Wet vs. Dry Turning.

Furthermore, it may be that in the particular cutting operation considered, there is greater threat of environmental and health liabilities for wet machining. The well known negative health effects associated with cutting fluid mist, as well as the potential liabilities and difficulties often associated with coolant disposal, might very well create the potential for greater environmental and health expenditures in wet machining. Therefore, when the costs of wet and dry machining are relatively close, the threat of environmental and health costs may be an important determining factor in deciding whether or not to implement a wet or dry machining process.

CUTTING FLUID RECYCLING

The analysis discussed thus far has not included the implementation of a coolant recycling system. Cutting fluid recycling often involves the use of filtering, skimming, separation, and pasteurization techniques to remove contaminants such as tramp oils, metal fines, and bacteria. Without removal of such contaminants, cutting fluids can degrade, losing their ability to lubricate and transfer heat away from the workpiece. Such fluid maintenance and recycling methods readily require additional capital expenditures for equipment, labor, and other related costs, while at the same time extending the life of the coolant and reducing the costs associated with fluid purchase and disposal. Therefore, from an economic standpoint, coolant recycling has become a significant part of cutting fluid use throughout the metalworking industry, especially with the ever-increasing cost of coolant disposal [8].

In order to address this issue, a coolant recycling system was integrated with the cost analysis previously outlined. An individual, self-contained, coolant recycling system, designed for small sump sizes, was incorporated into the previously described cost model. Including the cutting fluid recycling system in the analysis adds a fixed equipment purchase cost of $11,500, while at the same time increasing the life of the coolant and decreasing coolant disposal and new fluid purchases. In fact, the manufacturer of this recycling system has presented data that suggests a decrease of 75-90% in the amount of spent coolant disposed of annually.

For the purpose of this case study, it was assumed that the use of this coolant reconditioning system would reduce the amount of annual fluid disposal by 80%. Under this assumption, the implementation of the coolant recycling system was found to cut the coolant related costs in half, reducing them from 20% to 10% of the total machining cost. Of this 10%, 5% includes coolant-related equipment and maintenance costs, 3% includes the cost of new fluid purchases, and 2% includes costs of disposal. Therefore, cutting fluid recycling can have an important economic impact. In this case, the implementation of a fluid recycling system brought about a significant decrease in wet machining costs. In addition, the payback period for the coolant recycling system was calculated to be just over one year. Table 3 compares the changes in cost as a result of implementing the fluid recycling system.

	Without Recycling	With Recycling	Dry
Cost ($/part)	$20.23	$20.02	$19.87

Table 3: Economic Implications of Fluid Recycling.

AIR QUALITY

As previously mentioned, the generation of aerosols, which are typically defined as solid or liquid particles suspended in the air, can be very important, especially from a worker health perspective. This is of particular significance in machining processes, since cutting operations, such as turning, milling, drilling, and grinding often produce significant quantities of suspended metallic dust and/or cutting fluid mist.

The aerosol that has received the greatest attention in recent years is cutting fluid mist. The small droplets that form cutting fluid mist can conceivably be produced by two mechanisms: vaporization/condensation and atomization [9]. Atomization is the process by which mechanical energy is converted into droplet surface energy. Vaporization takes place as a result of the heat generated in the work zone. This heat, which is transferred to the fluid, may raise its temperature sufficiently above the saturation temperature, resulting in boiling and vapor production at the solid-liquid interface. This vapor may then condense around spontaneously generated liquid nuclei or other foreign particles to form mist.

EXPERIMENTAL PROCEDURE

In order to compare and contrast the aerosols generated as a result of wet and dry cutting, machining experiments were conducted in both the presence and absence of cutting fluid by turning cylinders of ISO R185 Grade 20 gray cast iron with a CNC Lathe. The machining area of the lathe was enclosed on all sides, with access to the spindle and turret provided via a sliding door. Sampling tubes were inserted through a small hole in the side of the machine enclosure and positioned roughly 0.6 meters above the cutting process, in order to measure the mass concentrations and particle size distributions of the aerosols produced during both wet and dry machining.

Although measuring outside the enclosure is more relevant to worker exposure and health, a sample location within the enclosure provides a more controlled environment and is better suited for studying the important factors and trends in aerosol generation. Mass concentration and particle size measurements recorded outside the enclosure may often be affected by the size of the room, the airflow patterns within the surrounding area, and the existence of other manufacturing operations within close proximity. The exact sampling location was selected, in large part, due to the limited space inside the enclosure, and was placed as far as possible above the cutting process, so as to prevent large fluid droplets and metal chips from disrupting the measurement process.

In the wet turning experiments, cutting fluid was applied at a flow rate of 12 L/min via a nozzle positioned directly below the cutting tool and directed toward the tool tip. The cutting fluid chosen, a synthetic metalworking fluid typically used with alloy steels and cast irons, was mixed with tap water to form a 5% solution. Cutting speed, feed, and depth of cut were identified as factors to be studied using a two level factorial design. The high and low levels of cutting conditions were selected to correspond to the range of values typically seen in the turning of cast iron and are listed in Table 4.

	Low Level	High Level
Cutting Speed (v_c)	152 m/min	381 m/min
Feed (f)	0.15 mm/rev	0.38 mm/rev
Depth of Cut (a_p)	0.5 mm	2.5 mm

Table 4: Cutting Conditions.

Also, in an effort to determine the role atomization plays in overall mist production, an additional set of tests were conducted in which fluid was applied to a rotating workpiece in the absence of cutting. These tests were identical to the wet turning experiments except that rather than cutting, the tool instead followed a path just beyond the outer edge of the rotating workpiece. These atomization tests were conducted at the same high and low levels of speed used in the cutting experiments so that comparisons could be made between the cutting fluid mist produced as a result of machining and that produced solely due to atomization.

The experimental procedure for both wet and dry cutting followed a randomized 2^3 factorial design. In all instances a replication was completed, giving a total of 16 tests for wet machining and 16 tests for dry. Each test involved the simultaneous measurement of the mass concentration and the particle size distribution before, during, and after the cutting process. A continuous measurement of the mass concentration was initiated 2 minutes prior to actual cutting, continued through the 2-3 minute cutting operation itself, and for 4 minutes following the machining process. In this manner the ambient condition before cutting, the cutting process itself, and the aerosol settling and dispersal following cutting could be sufficiently monitored. The particle size distribution was recorded over consecutive 30 second time intervals so that the change in the particle size distribution over time could be studied. Aside from the absence of cutting, the atomization tests were conducted in a similar manner, with a replication performed at each level of cutting speed, giving a total of 4 tests.

The mass concentration was measured using a laser photometer, which is essentially an aerosol monitor that determines mass concentrations in real time via the light scattering principle. The laser photometer was used to measure the concentration of particles with aerodynamic diameters between 0.1 and 10 microns. The particle size characteristics were measured via a particle sizing spectrometer, which uses the time-of-flight principle to measure particles between 0.5 and 20 microns.

AIR QUALITY RESULTS

In order to compare the aerosol mass concentrations recorded during wet and dry turning over equal time intervals, the mass concentration measurements were averaged over the first 90 seconds of the cutting process and used as the response variable in an analysis of variance. For the conditions considered, the analysis of variance revealed that all of the factors and interactions considered were significant in generating both dust and mist. As expected, the amount of mist and dust produced was found to increase with increasing values of speed, feed, and depth of cut.

In the case of both wet and dry machining the highest concentration of aerosol corresponded with the highest material removal rate. In fact, the least squares fit of the mass concentration data shown in Figure 4, does suggest a linear relationship between mist concentration and material removal rate. This linear relationship is most likely a product of the vaporization/condensation mechanism. Because the energy or power used in the material removal process is, in large part, converted to heat, the higher the material removal rate, the greater the amount of heat generated. Upon contact with the applied cutting fluid, greater amounts of heat likely serve to increase the amount of fluid that is vaporized, and consequently the amount of cutting fluid mist.

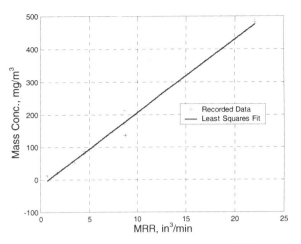

Figure 4: Average Concentration of Mist as a Function of Material Removal Rate.

Wet turning was found to produce substantially higher aerosol concentrations than dry turning. As illustrated in Figure 5, the amount of aerosol generated in wet machining, depending on cutting conditions, was found to be anywhere from 12 to 80 times greater than that produced in dry machining. In fact, as shown in Figure 5, the difference in aerosol concentration was so great that a log scale was necessary in comparing the aerosols generated in wet and dry turning at the following cutting conditions: speed = 152 m/min, feed = 0.38 mm/rev, and depth of cut = 2.5 mm.

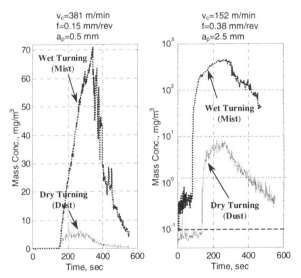

Figure 5: Aerosol Concentrations in Wet & Dry Turning.

The particle sizes generated for each aerosol were also found to be dependent upon cutting conditions. By comparing the particle size distributions at various cutting conditions, a number of qualitative observations could be made. In the case of dry machining a distinct difference was noticed between particle size distributions at different cutting speeds. At the high level of cutting speed (381 m/min) there were a larger number of particles generated in the 1-4 micron size range than at the lower speed (152 m/min). Unlike dry machining, the particle size distributions for cutting fluid mist showed much less variation over the range of cutting conditions, with the majority of particles generated being below 1 micron. As illustrated in Figure 6, the particle size distributions for cast iron dust and fluid mist were similar at low speeds, but at high speeds were found to be very different.

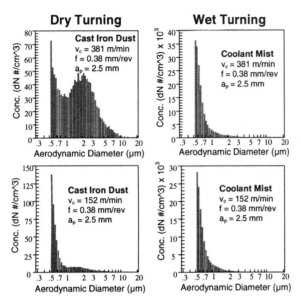

Figure 6: Size Distributions for Dust and Mist.

Slight differences were also observed among particle size distributions at different cutting conditions in wet machining. At low material removal rates the percentage of particles in the 1 – 2 micron size range was somewhat greater than at high material removal rates. This is likely due to the role that atomization plays at lower material removal rates. In general, the atomization of cutting fluid results in a particle size distribution with a greater percentage of larger particles (i.e., > 1 micron). The smaller amount of heat generated at low material removal rates results in less cutting fluid being generated via the vaporization/condensation mechanism. Therefore atomization plays a larger role in mist generation during cutting processes operated at low material removal rates. This may be why a greater percentage of larger particles are evident at lower material removal rates. Particle size distributions for both cutting and atomization are depicted in Figure 7.

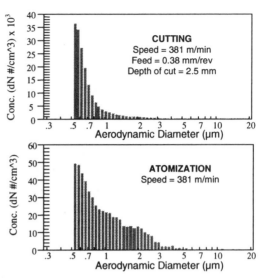

Figure 7: Distributions for Cutting and Atomization.

The role atomization plays in overall mist production can be understood more clearly by comparing the mass concentration measurements taken during the cutting process versus those measurements recorded exclusively during atomization. As seen in Figure 8, the concentration of coolant mist during cutting varied from 14 to more than 400 times that measured during atomization alone. Consequently, in such machining operations the vaporization/condensation mechanism of mist formation is responsible for the great majority of the mist produced, even for relatively moderate material removal rates.

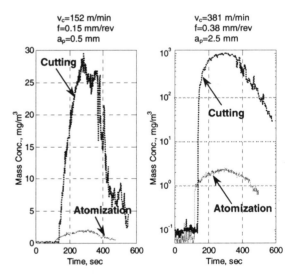

Figure 8: Cutting vs. Atomization.

AEROSOL COMPOSITION

The mass concentration and particle size were selected as the parameters for comparing the aerosols generated in wet and dry turning, primarily because they are the metrics used in the standards and regulations established by the Occupational Safety and Health Administration (OSHA) and other governing bodies involved in regulating particulate exposure. However, mass concentration and particle size alone are not necessarily indicative of the real health hazards individually associated with wet and dry machining. An aerosol's propensity to cause health problems depends not only on the amount and size of the particulate matter produced, but also on the composition and physical state of the aerosol. The toxicity and carcinogenity of the chemical components that make up the workpiece material, the tool material, and the cutting fluid, play a significant role in determining the true health risks of aerosol exposure. The growth of disease causing organisms such as bacteria and fungi within the cutting fluid can also be responsible for ill health effects [10]. Furthermore, since cutting fluids are typically diluted with large amounts of water, the mists resulting from their use may consist primarily of water. Therefore, while the mist concentrations reported in wet machining might be very high, the actual amount of exposure to potentially harmful aerosols may be significantly lower. Thus aerosol composition may be paramount, especially when comparing worker exposures in wet and dry machining.

PRODUCT QUALITY

Efficient process performance and favorable product quality are essential to any profit-oriented machining operation. If the parts produced do not meet the required size, shape, and finish they can not be sold to generate revenue. Therefore, the effect cutting fluid utilization has on these issues must be understood when contemplating whether to implement a wet or dry machining process. If the role of cutting fluid is found to be fundamental to machine tool performance and/or product quality, either coolant must be used or the functions of cutting fluid must be met by alternative means. This could involve alterations such as adjustments to the machining parameters, the purchase of new machine tools or related attachments, the use of different material tool types or geometries, or the development of new technologies, all of which can influence the machining economics or result in additional capital expenditures.

This research focuses on the effect of cutting fluid application on the product quality and process performance associated with the turning of gray cast iron. Cutting forces, surface finish, and dimensional error were recorded, along with the air quality parameters measured in the wet and dry turning experiments discussed in the previous section. The resulting data was analyzed and used to determine the physical role cutting fluid plays in this specific turning example.

EXPERIMENTAL PROCEDURE

The cutting forces, surface roughness, and dimensional error were measured in both wet and dry turning. For analysis purposes the data was combined into a 2^4 factorial design, with the four factors being speed, feed, depth of cut, and coolant. The levels of speed, feed, and depth of cut were as previously specified in the air quality experiments. The low level for coolant corresponded to dry machining, whereas the high level corresponded to wet machining, with the coolant applied at a rate of 12 L/min.

During the cutting process of each test, the cutting forces, which included the cutting, feed, and thrust forces, were measured using a strain gauge type dynamometer. The force signals were sampled and recorded by a data acquisition board mounted in a PC. Following each test, a profilometer was used to record a minimum of 5 surface roughness (Ra) readings at locations randomly selected around the circumference of the workpiece. Close attention was paid to the variation between measurements. If little variation was observed, the 5 readings were considered sufficient. However, if substantial variation was evident, additional measurements were taken. The machined part diameter was also measured, from which the dimensional error was calculated. This procedure was followed for both wet and dry machining, thus giving a total of 32 measurements for the cutting forces, surface roughness, and dimensional error.

PRODUCT QUALITY RESULTS

While some differences in the cutting forces, dimensional error, and surface finish were noted throughout the turning tests, the application of cutting fluid was not found to produce large improvements in product quality or process performance. In fact, feed was the only factor found to have a significant influence on the surface finish of the workpiece. The surface roughness was found to increase noticeably with increasing values of feed, while speed, depth of cut, and coolant application were not found to have a significant effect. Therefore under the conditions specified, wet machining was not found to have an advantage over dry machining, from a surface finish perspective. Furthermore, none of the factors considered (i.e., speed, feed, depth of cut, and coolant) were found to play a statistically significant role (significance level, $\alpha=0.05$) in the dimensional accuracy achieved. More importantly, coolant was not found to promote decreases in dimensional error, thus providing no apparent advantages in accuracy over dry machining.

The cutting force was found to be significantly affected by speed, feed, and depth of cut, while coolant use was not found to be statistically significant. However, while coolant use was found to have no significant effect on the cutting force, this was not true of the feed and thrust forces. In fact, in some cutting conditions, particularly those at the highest material removal rate, feed and thrust forces were found to be reduced by as much as 30%, as a result of coolant application. Therefore the findings do suggest that wet machining does provide a small advantage over dry machining in the area of force reduction. Figure 9 illustrates the lower feed and thrust forces recorded at the highest material removal rate.

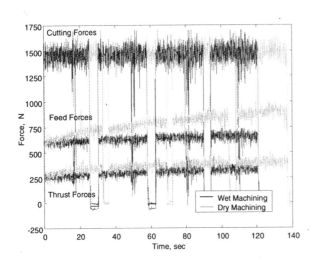

Figure 9: Forces at (v_c=381 m/min, f=0.38 mm/rev, a_p=2.5 mm) for Wet & Dry Turning.

SUMMARY AND CONCLUSIONS

The decision of whether to implement wet or dry machining for a particular machining process depends on a number of factors, most, if not all of which, involve manufacturing economics. After all, the goal of virtually any machining facility is to maximize company profits. A cost model was developed for the purpose of comparing the unit costs associated with wet and dry machining. Although the model was developed with turning in mind, it can, with very few changes, be used to compare the unit costs and annual profit for virtually any machining operation. The framework of the model can also be utilized in performing economic comparisons of wet and dry machining lines by applying the unit cost model to each individual operation within the machining line considered.

Aside from the traditional machining costs such as tooling, labor, power, and equipment costs, product quality and environmental/health issues can also have serious economic implications. This research has integrated these traditional machining costs with the potential environmental/health liabilities and product quality related issues for the process of turning gray cast iron.

For the specific conditions considered, the machining costs in dry turning were found to be noticeably lower than those in wet turning. The underlying reason behind these cost differences was the additional capital expenditures associated with coolant use. In fact, for wet machining, without the use of a coolant recycling system, coolant-related costs were found to account for as much as 20% of the total machining costs, whereas tool costs accounted for only 3%. In addition, the implementation of a coolant recycling system was found to reduce coolant-related costs in wet machining by over 50%.

In the experiments focused on, air quality, speed, feed, and depth of cut were all found to have statistically significant effects on the amount of mist and dust produced in wet and dry turning. The aerosol mass concentration in both wet and dry cutting was found to increase with increasing values of speed, feed, and depth of cut, with the greatest concentration levels recorded at the highest material removal rates. The air quality experiments also revealed that the concentrations of suspended cutting fluid mist far outweighed that of the cast iron dust generated in dry turning. In fact, wet turning was found to produce particulate mass concentrations 12-80 times that recorded during dry turning. Furthermore, the mist produced in wet cutting was found, in many cases, to consist of much smaller particle sizes and settled out of the air at a much slower rate than did the dust created in dry cutting. Therefore, composition aside, wet machining has a much greater propensity to cause adverse health effects in this turning operation. Not only is the worker exposed to a greater quantity of particulate matter, but the smaller sized mist particles remain suspended in the working environment for a longer duration and can be more easily inhaled into the deep portion of the lungs.

In trying to suppress mist generation, it is vital to understand the mechanism by which the mist is created. A comparison of the experimental results revealed that the vaporization/condensation mechanism contributes much more to mist generation than does atomization. In fact, for the conditions under study, the cutting process was found to produce mist concentrations 14-400 times greater than that generated via atomization alone.

However, regarding worker exposure and health, the factor that must not be overlooked is the composition of the suspended particulate matter. Cutting fluids used in machining are typically diluted with large amounts of water. As a result, the mists generated may consist primarily of water. Thus, while the mass concentrations reported in wet machining might be very high, the actual amount of exposure to harmful airborne particulate may be significantly less. Therefore, in comparing the air quality associated with wet and dry machining, the composition of the particulate matter may be of ultimate importance. The toxicity, carcinogenity, and biological affinity of the airborne particles may play a significant role in determining the likelihood of a particular machining process to cause adverse health effects.

The product quality experiments did not show any large advantages or fundamental requirements of cutting fluid use for the turning operation studied. Cutting fluid was not found to have a significant effect on either dimensional error or surface finish, but was, however, observed to wash away and remove chips from the machine tool's slideways and other moving parts. Cutting fluid application was also found to reduce both the feed and thrust forces by as much as 30% at the highest material removal rates, although this might be a product of the increased heat removal and tool life and may have little practical importance.

In estimating the true cost associated with a particular machining process, all related issues such as machining costs, product quality, worker health, and environmental liability must be reviewed. For the specific case study considered, final calculations revealed that the production costs in dry turning were noticeably less than those in wet turning. Also, considering the limited product quality benefits, the large aerosol concentrations, and the potential environmental liabilities associated with cutting fluid use, dry turning appears to offer a substantial advantage in this case. Therefore, by considering the traditional machining costs, as well as the relevant product quality, worker health, and environmental issues, a sound economic decision regarding whether or not to use cutting fluid can be effectively made.

ACKNOWLEDGMENTS

This research was supported, in part, by the U.S. Department of Education's Graduate Assistance in Areas of National Need (GAANN) Program, the National Center for Clean Industrial Treatment Technologies (CenCITT), and the Ford Motor Company. Kennametal, Caterpillar, and A.W. Chesterton are gratefully acknowledged for their material and equipment donations.

REFERENCES

[1] Hands, D., Sheehan, M. J., Wong, B., Lick, H. B., 1996, Comparison of Metalworking Fluid Mist Exposures from Machining with Different Levels of Machine Enclosure, American Industrial Hygiene Association Journal, 57/12:1173-1178.

[2] Mackerer, C. R., 1989, Health Effects of Oil Mists: A Brief Review, Toxicology and Industrial Health, 5:429-440.

[3] McClellan, R.O., Miller, F.J., 1997, An Overview of EPA's Proposed Revision of the Particulate Matter Standard, Chemical Industry Institute of Toxicology Activities, 17/4:1-24.

[4] Wickham, T., "UAW Pushes for Standards on Fluids," The Flint Journal, July 26, 1999.

[5] NIOSH, 1998, What You Need to Know About Occupational Exposure to Metalworking Fluids, DHHS (NIOSH) Publication No. 98-116.

[6] Yu, J.J., W.K. Chen, M.C. Mei, and M.H. Tsou, "Pulmonary Siderosis: Report of a Case," Journal of Formos Medical Association, Supplement 4, 1993, pp. 258-261.

[7] Kress, D., "Dry Cutting With Finish Machining Tools," Industrial Diamond Review, Vol. 57, No. 3, 1997, pp. 81-82.

[8] Sluhan, W. A., " Don't Recycle – Keep Your Coolant," American Machinist, Vol. 137, No. 10, 1993, pp. 53-55.

[9] Yue, Y., Olson, W.W., Sutherland, J.W., 1996, Cutting Fluid Mist Formation via Atomization Mechanisms, Proc. of Symp. on Design for Manufacturing and Assembly, ASME Bound Volume – DE 89:37-46.

[10] Thorne, P. S., DeKoster, J.A., Subramanian, P., 1996, Environmental Assessment of Aerosols, Bioaerosols, and Airborne Endotoxins in a Machining Plant, American Industrial Hygiene Association Journal, 57/12:1163-1167.

GMBOND™ Process: An Environmentally Friendly Sand Binder System

Scott R. Giese and Gerard Thiel
University Of Northern Iowa Metal Casting Center

Richard M. Herried and Jeremy D. Eastman
Hormel Foods Corp.

ABSTRACT

Automotive engineers are challenged with increasing fuel economy in transportation vehicles by reducing weight. Aluminum castings are replacing cast iron components as one way to reduce weight in cars. Many of the aluminum castings produced for automobiles today are made with a sand core to form the internal cavity of the automotive component. Currently, the most popular choice of sand binder system in core making is the phenolic urethane cold-box binder system. This system, however, was not designed for use at aluminum pouring temperatures.

The collapsibility of a phenolic urethane cold-box core is not sufficient for expedient shakeout in an aluminum casting. Because of this, many aluminum castings must undergo secondary core removal processes. This reduced shakeout effectiveness limits the designer in the casting geometries available and adds cost to the part. The protein-based binder system was designed to replace current binder systems as an environmentally friendly sand binder and to improve the shakeout characteristics of the sand core.

Detailed benefits of the protein-based binder process include excellent shakeout effectiveness, recyclability, and cost savings because of the reduced solid waste and labor expenditures. More applications for aluminum casting replacements will become available as the awareness of this new process grows. Dissemination of the state of technology of the protein-based binder process to the automotive engineer would allow for more timely advancement of casting design and metal casting technology.

INTRODUCTION

In the early 1990's, researchers at General Motors began to evaluate many different materials that had binder-like qualities. Each material under consideration had to meet two criteria: (1) environmentally innocuous and (2) handling strength comparable to the phenolic urethane cold-box (PUCB) process. Purified proteins from the agricultural industry were investigated as possible candidates that exhibited non-toxic degradability and can be synthesized with a variety of physical properties. A new binder system was developed that used a mixture of various molecular weight proteins intimately mixed with a metal oxide catalyst to enhance thermal degradation. In 1994 a patent[1] was issued to inventors of this new binder system, trademarked GMBOND™ Sand Binder. Subsequent core production developmental work was performed at the GM Powertrain Group Advanced Developmental Lab. Consequently, a patent[2] was issued in 1996, which detailed the protein-based binder core production process. After years of developmental work, the protein-based binder system was introduced to the foundry industry. Since then, casting trials have been performed to evaluate casting quality and enhance the sand binder's effectiveness.

Unique production and environmental challenges for the 21st century are confronting the metal casting industry. New environmental regulations are planned to be enacted with complete compliance by the foundry industry in the next several years. As these laws are implemented, broad range impacts of these regulations will not only affect the internal environmental and production expenses to a foundry but also disposal costs. In addition to stricter environmental regulations, disposal costs will continue to increase with no foreseeable reduction in the future. Foundries need to find innovative technologies and materials to reduce and prevent the production of solid waste emissions and create closed recycling loops of foundry materials.

The commercial availability of protein-based binder system to the foundry industry addresses a variety of issues confronting the metal casting industry. In *American Metalcasting Industry Technology Roadmap*, development of "new processes or materials to reduce or prevent the production of foundry wastes" was identified as a high research priority in the environmental technology area[3]. Areas identified include 95% recycling of waste materials, energy efficient sand reclamation technologies, and exploitation of new materials to reduce or prevent production waste. This paper discusses the advantages and processing parameters of the protein-based binder system and how they address the environmental initiatives endorsed by the metal casting industry.

THE BINDER

The protein-based binder system is available in a dry fine powder. It is neither flammable nor reactive. The binder is completely soluble in water and has a specific gravity greater than water. Raw materials for the binder are derived from renewable natural sources and are environmentally benign. These raw materials are purified and processed to give the protein-based binder characteristics that allow it to be used as a sand binder. The protein-based binder is a combination of various types of polypeptide molecules. Thermal degradation of the binder is improved with small additions of iron oxide catalyst. A small amount of preservative is also added to improve bench life for several days while the mixed core sand is wetted. The preservative in the protein-based binder is safe to animals and humans if the binder were ever ingested.

One of the most important features of the protein-based binder is its water solubility. The bonding mechanism is completed by dehydration of the wet sand mixture. The bond is not covalently cross-linked as in organic binder systems and can degrade thermally at lower casting temperatures. The polypeptide molecules form a partially crystalline structure as water is removed from the core. Although the bonding mechanism for the protein-based binder is reversible with the addition of water, when dry the partially crystalline structure does not melt at high temperatures.

THE GMBOND™ PROCESS

SAND PREPARATION

Applying binder onto sand is accomplished by combining sand, binder, and water while mixing as illustrated in Figure 1. To begin coating of the protein-based binder, the dry binder is added to hot sand during mixing followed by the addition of water and mulled until dry. During mulling, the sand must retain enough heat to evaporate the water. The final, dry coated sand has the appearance of clean, new sand.

The amount of binder used during coating will determine the water addition during conditioning. This ratio determines the relative moisture content of the sand mixture. Too much sand moisture will affect flowability and drying time. Using the smallest amount of water possible during conditioning will make it easier to give the core sufficient handling strength in an efficient manner.

Coating Sand with GMBOND™ Sand Binder

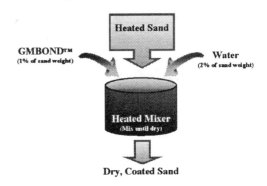

Figure 1 - Sand preparation of sand for the protein-based binder process.

SAND CONDITIONING

Prior to blowing the core, the cool sand is conditioned with water and mixed to evenly hydrate each coated sand grain. The sand conditioning process is illustrated in Figure 2. The dry coated sand is first cooled to remove excessive heat to impart better flowability properties to the hydrated sand particle. Water is then added to an amount that hydrates the sand particles. Excessive water will have a detrimental affect on the core production process.

Rehydrating GMBOND™ Coated Sand

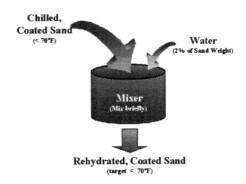

Figure 2 - Sand conditioning of sand for the protein-based binder process.

Blowing Cores with GMBOND™

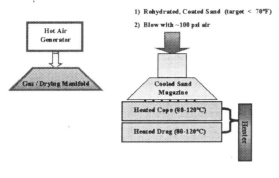

1) Rehydrated, Coated Sand (target < 70°F)
2) Blow with ~100 psi air

Activating GMBOND™

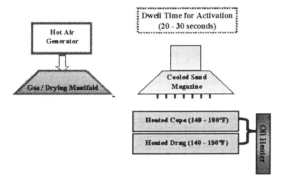

Dwell Time for Activation
(20 - 30 seconds)

Hardening GMBOND™ Cores

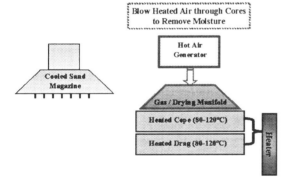

Blow Heated Air through Cores
to Remove Moisture

Figure 3 - Schematic illustration of the core production methodology used to create protein-based binder cores. Sand is initially blown into a heated core box and allowed to activate for a predetermined time. The hardening of the binder is completed when dry air is purged through the core. Afterwards, the core is removed and transferred to production.

CORE PRODUCTION

Core forming operations are completed by the controlled dehydration of the sand mixture within a heated core box. Figure 3 shows a schematic illustration of the core production process for the protein-based binder system. First, conditioned sand is blown into a core box and is termed as the *blow cycle*. As the sand grains are packed tightly into the heated core box cavity, the binder

from adjacent sand grains begins to coalesce at the contact points, allowing the binder to activate. This process is termed as the *activation cycle*. Dry air is blown into the core box to impart core strength by the dehydration or curing of the core. The introduction of dry air is called the *purge or drying cycle*.

PROCESS CONSIDERATIONS

The protein-based binder process differs from other sand binder systems for core production requirements. The core machine for the protein-based binder contrasts from a cold-box core machine in two ways: the addition of a heated core box and removal of the amine generator. Since the protein-based binder is a hot-box type core production process, high production runs require a core box material having good thermal conductivity and wear resistance. Typical tooling candidates are aluminum and gray iron. Prototype tooling such as high-temperature plastic materials can be used provided the plastic tooling has good thermal conductivity. An air dryer, with a heater and moisture trap, is substituted for the amine generator. Adequate curing of the binder requires sufficient removal of water.

The cores are typically stripped with some residual water remaining within the center of the core. The binding mechanism of the protein-based binder system operates similarly to the acid catalyzed furan binder system by forming a cured skin at the core box wall and drying inward. The drying cycle will purge most of the moisture from the core and the heat retained during the activation cycle will provide a thorough cure as the remainder of the core dries after stripping from the core box. Though some larger cores might contain residual water in the center resulting in "soft" spots, cores can be handled out of the box and used immediately.

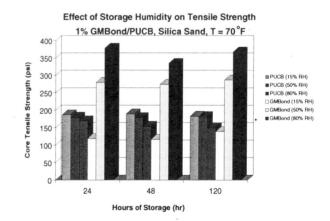

Figure 4 - Tensile strength properties as a function of temperature and humidity. The protein-based binder properties were compared to the phenolic urethane cold-box (PUCB) process at 1% binder addition. The protein-based binder stored at 80% relative humidity (RH) was redried in an oven at 105°C. The redried sample increased to 311 psi.

39

Most of the strength is obtained within the first few hours of manufacture as shown in Figure 4. The effect of humidity has minimal influence on the tensile properties during the first fours hours after curing regardless of the humidity. Significant loss of strength occurs over time (greater than 4 hours) and is more pronounced in higher humidity environments. Typically, this effect would significantly affect the quality of the casting. However, tensile strength can be recovered if cores are placed into a core oven for a short period time to remove any accumulated moisture as indicated in Figure 4.

For a foundry, proper process control include a monitoring of the conditioned sand moisture and temperature, air pressure for blowing and purging, core box temperature, and venting. These process variables will determine the total cycle time and core properties. The activation and drying cycle times are closely monitored and adjusted for each type of core geometry. An intricate core having a high surface area to mass ratio will activate and dry faster than a bulky core having low surface area to mass ratio. For any core geometry, activation time and purge time is adjusted to permit migration of the binder to the contact points.

The full strength of the protein-based binder is not necessary when producing high production castings. For handling, cores do not have to be completely dried since a strong, hard shell develops during the curing cycle. Cores are stripped with no more than 15% retained water of the total amount added. Strength proceeds to increase after the core is removed from the core box as it continues to dry. However, certain applications requiring high strength would be beneficial for selected casting process. Full strength is achieved when water is completely removed.

As with many binder systems, core refractory coatings can be used to improve the surface finish and prevent sand-related defects. Water based coatings have been successfully applied to protein-based binder cores. Drying of low-solid core coatings require immediate reheating to retard the absorption of water. High solid water-based coatings can be flow coated or spray coated without reheating. Alcohol based core coatings can be used without any degradation of the binder though environmental concerns arise with the use of these core coatings. A process[3] was developed and patented by the GM researchers for core coating the protein-based binder cores. It involves waterproofing the core prior to core coating with a water based refractory coating. An alternative method was developed that involves the powder coating of a refractory material, which would be used instead of the water based coating. Additionally, selective coating of critical areas is possible with this technique.

The type of molding aggregate does not effect the bonding mechanism and properties of the protein-based binder. Most organic binder systems are limited to certain molding aggregates depending on the sand chemistry and acid demand value. For example, olivine sand is not compatible to phenolic urethane binder systems. Table 1 shows the typical tensile strength for a variety of molding aggregates using the protein-based binder. Tensile strength for silica, lake and zircon sand shown in Table 1 that are comparable to strengths produced with the phenolic urethane binder system. The other molding aggregates listed are not compatible or are difficult to use with phenolic urethane. Binder additions of 0.75% to 1.25% based on sand weight have been used to produce cores with high strength.

Table 1 - Tensile strength of GMBOND for a variety of molding aggregates.

Type of Sand	GFN	GMBOND (based on ρ_{sand})	Tensile (psi)
Silica	35	1.0%	230
Lake	70	1.0%	270
Olivine	100	1.0%	160
Zircon	110	0.6%	280
Chromite	50	0.6%	230
Mullite	60	1.0%	230

PRODUCTION TRIALS

Casting trials were performed at Teksid S. p. A. in Carmagnola, Italy on an aluminum suspension arm part[5]. The suspension part was converted from a steel weldment. It employed a square tube design with relatively thin walls, 3 – 4mm, which improved strength and reduced weight. This part had a high surface area to weight ratio, which does not provide much heat energy to thermally degrade the binder in the sand core. Since the suspension arm has safety requirements, no intensive shakeout could be preformed without the likelihood of damaging the casting. Originally, it was planned that a furan hot-box core would be employed as the core production process. To remove the core from the suspension arm casting, an 11-hour heat treat cycle was planned before heating treating for metallurgical properties.

A casting trial was performed using a protein-based binder core with the low-pressure semi-permanent mold process to evaluate shakeout and dimensional accuracy. The production trial work was performed on a cold-box core machine that was modified to cool sand before mixing and blowing. Dry shakeout was accomplished by blowing compressed air through the casting after cool the casting for 30 minutes in ambient air temperatures. Wet shakeout was investigated by directly quenching the part after casting. Dimensional accuracy was verified and showed no thermoplastic deformation. The casting surface finish was excellent despite a 45% lower binder content than the proposed furan hot-box core. After the

2nd day of trials, core scrap was reduced to less than 2% with a 60-second production time.

RECYCLING & RECLAMATION

Any scrap core made with the protein-based binder can be recycled back into the core production system without having to add any additional binder. This includes any remaining sand that will not be used before it will dry, i.e. sand left in sand hopper at the end of the production shift. Tests have shown that cores can be made with little loss of strength or surface hardness. Additionally, experiments have been performed on thermally degraded protein-based coated sand and then recoated with the protein-based binder as shown in Figure 5. Slight loss in tensile properties were observed but within production requirements.

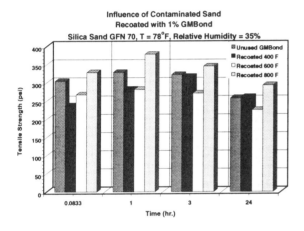

Figure 5 - The effect of recoating GMBOND™ sand exposed to elevated temperatures.

Sand reclamation can be accomplished by one of three different methods: thermal, mechanical, or wet. The protein-based binder thermally decomposes at reasonably low temperatures and may be accomplished relatively quickly. Generally, thermal reclamation processes use more energy and contribute greenhouse gas emissions and other air pollutants. Therefore, mechanical reclamation and wet reclamation are the most environmentally friendly processes available to reclaim the protein-based bonded sand because the particulate residue from reclamation is non-toxic. This material could readily be added to a landfill.

Table 2 - Loss-on-Ignition of GMBOND™ coated sand.

Heat Treament (One hour treatment time)	Loss-on-Ignition (1400°F @ I hour)
Untreated	1.17%
400°F	0.77%
600°F	0.38%
800°F	0.00%

Reusing either thermally or mechanically reclaimed sand can be done readily with the protein-based binder system. The heat treatments given the protein-based coated sand in Figure 5 range from slight binder decomposition (400°F) to moderate decomposition (600°F) to complete thermal reclamation (800°F). The loss on Ignition values for each of these treatments before coating is shown in Table 2. Typical mechanically reclaimed sand from the casting process could contain sand with all of these treatments depending on the sand to metal ratio. Recoating this sand produce tensile strengths that meet production requirements.

ENVIRONMENTAL BENEFITS

The protein-based binder is safe to use and does not use toxic chemicals or chemical scrubbers. There is relatively no odor during coremaking. Employees do not need to use special personal protective equipment when working with the binder. Water is used to clean tooling. Improved shakeout and water solubility will eliminate dusts generated during intense core removal processes. The protein-based binder is neither corrosive nor flammable. No special containers are necessary for storage or transportation.

The ability to easily reclaim and reuse sand will reduce the waste generated in the foundry. Spent sand can be easily reclaimed without creating additional hazards. Flawed cores can be reused and don't have to be thrown away. If sand were to be disposed, it would not be toxic.

CASTING BENEFITS

Cores made with the protein-based binder produce castings with excellent results. The binder can produce castings with greater dimensional accuracy, excellent surface finish without having to perform secondary core removal processes. Castings made with the protein-based binder have low core related scrap problems. The binder breaks down at lower temperatures with the aid of the catalyst. The gas evolution is predictable and somewhat accelerated as compared to other binder systems. The amount of total gas evolved during casting may be decreased since less binder can be used to produce cores of adequate strength.

One of the most important features of this binder system is the simplified core removal of aluminum and magnesium casting. The heat at those pouring temperatures will easily degrade the binder sufficiently to allow for dry shakeout of the core, without having to perform thermal core removal. Most of the time the core sand will flow relatively freely from the casting. Another important feature is the ability to do a wet core removal. Simply submersing the casting in water will dissolve any binder that has not degraded thermally. Any remaining sand can be flushed away from the casting.

CONCLUSION

The benign nature and physical characteristics of the protein-based binder offer many environmental and casting advantages. This new core binder eliminates the use of toxic chemicals and disposal of toxic waste. Castings exhibiting shakeout problems can now benefit from a binder system especially designed for greater collapsibility.

REFERENCES

1. *United States Patent 5,320,157, Expendable Core For Casting Processes.* Jun. 14, 1994. Jung-Sang Siak, Richard M. Schreck, Kush K. Shah.

2. *United States Patent 5,582,231, Sand Mold Member And Method.* Dec. 10, 1996. Jung-Sang Siak, William T. Whited, Mark A. Datte, Richard M. Schreck.

3. "Environmental Technology", *Metalcasting Industry Technology Roadmap*, U.S. Department of Energy, 1998, pp 41-49.

4. United States Patent 5,749,409, Method of Forming Refractory Coated Foundry Core. May 12, 1998. Jung-Sang Siak, Scott W. Biederman, William T. Whited, Mark A. Datte.

5. Ferrero, A., Badiali, M., Schreck, R., Siak, J. and Whited, W. "New Binder for Casting Cores: An Industrial Application to Safety Suspension Parts", SAE International Congress and Exposition, February 23-26, 1998.

Recovery of Waste Polystyrene Generated by Lost Foam Technology in the Automotive Industry

Jeremy Pletka and Jaroslaw Drelich
Michigan Technological Univ.

ABSTRACT

In the automotive industry, lost foam casting is a relatively new technology, which is gaining popularity among manufacturers. Lost foam casting is a process in which an expanded polystyrene pattern is formed into the shape of the part to be cast. More complex parts are fabricated by simply gluing several simple patterns together. The pattern is then coated with a refractory material consisting of a mineral mixture and binders. Finally, hot metal is poured into the pattern, evaporating the expanded polystyrene and taking shape of the coating shell. However, the automotive industry has observed that a significant number of these fabricated, coated patterns are damaged, or do not meet specifications prior to casting. These are not reusable and inevitably are landfilled. It is the goal of this project to develop a simple, reliable, and inexpensive technology to recover expanded polystyrene from the glue and coating constituents. This study includes an investigation of the size distributions of each component (polystyrene, coating, and glue) upon shredding, air separation/classification, and the effectiveness of impact milling. Separation testing has recovered as high as 96% of the polystyrene, while the level of contaminants did not exceed 5 wt% in the final product. These results indicate that a promising technological approach has been selected and that realization of a separation technology is feasible.

INTRODUCTION

Lost foam casting is a rapidly growing technology and is gaining confidence among manufacturers. A variety of benefits such as lower machining requirements on cast products and higher surface quality have led to an overall decrease in manufacturing costs and have bolstered its use in recent years. In 1997, 140,700 tons of aluminum, iron, and steel (of which aluminum is a majority) were cast using lost foam technology. A recent market study indicates that production will increase by 83% to 256,800 tons in 2000. Thus, as lost foam casting becomes more popular among manufacturers, the issue of waste disposal becomes a more pertinent issue [2,3].

In the automotive industry, lost foam casting is a process in which expanded polystyrene (PS) molded in the shape of the part to be cast is replaced by hot metal. Complex parts are fabricated by simply gluing several simple pattern pieces together. The final pattern is then coated with a refractory material consisting of a mineral mixture and binders. Hot metal is then poured into the casting, replacing the expanded polystyrene, and taking shape of the coating shell.

A significant number of these casting patterns are rejected prior to casting due to defects, or to damage caused by handling. Damaged casting patterns are rejected from the casting process, compressed to minimize volume, and landfilled. Increasing landfill costs and, an increasing environmental awareness have precipitated initiatives to recover and recycle the expanded polystyrene portion of rejected casting patterns. Currently, no technology exists for the recycling of polystyrene from rejected casting patterns, and no prior attempts at developing a technology have been made.

The implementation of a recycling process for rejected casting patterns has a three-fold potential: reduction in landfill costs, reduction in the environmental impact of waste disposal, and reuse of a petroleum-based resource. However, these benefits cannot be realized if the recovery technology is not economically competitive; the technology must be simple, flexible, and inexpensive. In view of this, laboratory experiments were constrained to dry (without the use of water or liquids) and based alone on physical characteristics.

In the minerals industry, metals and industrial minerals are processed using two basic principals: liberation and separation. That is, the ore is milled to a specific size where the valuable material is distinct species, while the invaluable (gangue) material is a distinct particulate species in itself. The liberated valuables are separated from the gangue by the manipulation of some chemical or physical characteristics. These may include surface chemistry, density, magnetic properties, or various others. Almost all non-synthetic raw materials are produced in this manner.

These simple principals are easily applied to rejected casting patterns. This is because the constituents (expanded polystyrene, glue, and coating) are physically

Table 1 - Estimated composition of rejected casting patterns and other properties of expanded PS, coating, and glue.

Constituent	Weight %	Density (g/cm^3)	Behavior under Impact stressing
PS (Expanded)	35	0.02-0.05	Elastic/visco-elastic deformation
Coating	60	2.65	Brittle fracture
Glue	5	0.92	Brittle fracture

distinct, which means that they, like metallic ores, can be reduced to an appropriate size for liberation of the PS from the coating and glue. Additionally, the density differences between the components (shown in Table 1) suggests that a separation process based on density will be successful.

EXPERIMENTAL

The experimental work performed for the development of a process technology for the recovery of polystyrene from rejected casting patterns consists of size reduction, measurement and analysis of particulate size distributions, screening, impact comminution, and density based separation. Experimental work was performed on samples obtained from casting patterns rejected from GM powertrain. Table 1 shows the initial composition of the casting patterns in weight percent.

Sample analysis was performed with an ashing technique. Representative samples were placed in enclosed crucibles and subjected to high temperatures, which burn off all volatile material (glue and PS). The remaining coating could then be weighed and a weight percent coating calculated. Note that this procedure does not distinguish between PS and glue.

SIZE DISTRIBUTIONS – The size distributions of coating, PS, and liberated glue from shredded casting patterns were determined individually. This was accomplished by sieving a 400-500g sample on a stack of sieves following a root-2 progression to 65 mesh on the Tyler scale. Due to the volume represented by a 400-500g sample, the material was sieved in smaller sized batches of about 40-80 grams depending on the range of size distribution. Each batch was sieved for 20 minutes. Following this procedure each size fraction was subjected to a two-stage sink/float procedure.

A two-stage sink/float procedure is used to minimize the contact of coating with the low density organic liquid ethanol. Ethanol dissolves a portion of the coating, presumably the binder, leading to a weight loss of up to 4%. However, since shredding releases most of the coating (>95%) most of it can initially be separated in a water medium. Therefore by minimizing the amount of

coating coming in contact with ethanol, and by minimizing the length of time in contact (about 10 minutes), the weight loss is insignificant. The first stage consists of immersing the material in a water bath followed by 2-5 minutes of gentle agitation. The dense material was then allowed to settle out for 5-10 minutes followed by removal of the floats. The floats consist of a mixture of PS and liberated glue, since glue has a density slightly less than that of water. The floats were dried, and subjected to a second float/sink stage in ethanol, which has a specific gravity of 0.79. The liberated glue would sink after following a similar procedure as mentioned above for float/sink water testing. The liberated glue was dried, weighed, and added to the coating rich water sinks for ash analysis, where the weight percent of liberated glue is determined. The floats were then ashed to reveal the amount of unliberated coating (coating still adhering to the PS particles). Note that at this point no distinction is made between PS and unliberated glue (glue that is still adhering to PS particles.

SHREDDING – Shredding of casting patterns was performed with a rotary style granulator, where cutting action was facilitated with a set of three knife blades. At the bottom of the cutting chamber a screen was placed to control the top size of the shredder product. Screens of 1, ¾, ½, and ¼ inch aperture diameter were used. Static generation during shredding caused PS particles to stick to the sides of the cutting chamber, and not report through the screen to the product bin. To reduce the amount of sticking material, a vacuum was applied to the product bin, generating negative pressure throughout the shredder. Each batch of material fed through the shredder weighed a minimum of 1000g to ensure a constant composition.

HAMMERMILLING – Impact comminution denotes the application of large stresses at high speeds to particulate material. A hammer mill, which utilizes impact-type breakage, consists of pivoted striking bars that impart the stress to the particles, and have a plate screen at the bottom of the mill to control particle size [5].

The purpose of these experiments was to determine to what extent the coating can be removed from coarse PS particles, and under what conditions does this occur. Of primary concern is the effect of residence time and mill RPM. Therefore, a series of tests were ran at 500, 1000, and 1500 mill RPM, and 30, 60, and 90 seconds of residence time. For statistical purposes each one of these conditions was performed twice, all in random order. All experiments were performed in a batch style on a representative 25g sample of coarse PS particles (+1/4-inch) from material shredded with the 1-inch screen.

It was observed that a significant portion of material tended to statically cling to the sides of the mill and escape comminution. In order to suspend all material in the mill, air was blown across the top, and down the backside of the mill. Air at the same flow rate was then drawn from the bottom of the mill through a plate screen, which held the material in the mill. The airflow reduced the amount of material that statically clung to the sides of the mill, but did not completely eliminate the problem. Only individual PS beads clung to the sides of the chamber. After milling, the material was separated by size with an 8-mesh screen.

DENSITY BASED SEPARATION – Separation of coating and glue from expanded polystyrene particles was facilitated with an air-gravity settling box, which is shown in Figure 1. The settling box consists of a 20" by 20" tunnel extending a total of 96" in length and is divided into two sections each 48" in length. The first section serves the purpose of not only housing the fan, but also to straighten the flow and provide a more uniform velocity profile. These are extremely important considerations for the design of the settling box. Preliminary experiments showed that expanded polystyrene is very sensitive to local disturbances (turbulence) in the air stream due to the extremely light density of polystyrene. Therefore, air streams with a high degree of turbulence tended to settle polystyrene almost at random distances, rather than at distances characteristic of the particles. To reduce turbulence a bladed fan was used. Bladed fans produce an air stream with less inherent turbulence than other types of blowers.

Figure 1 - Photograph of gravity settling box used in density separations.

Casting patterns shredded past the ¼" exit screen were subjected to separation in the settling box. Samples of 200g were fed to the settling box at an average feed rate of 3.3g/min, and an average air speed of 2.0m/s. The material settled into 7 separate sections: 6 equally spaced sections comprising of 8" lengths plus the material that overflowed the entire 48" section onto a table outside of the box. Each of the sections was then subjected to a stack of sieves ranging from 4-mesh to 20-mesh following a root-2 progression. Material losses were 1-2wt%.

RESULTS & DISCUSSION

A process, as shown in Figure 2, has been developed for the recovery of expanded polystyrene from rejected casting patterns. Rejected casting patterns consisting of expanded PS, glue, and coating are fed into a shredder which reduces the patterns into a particulate form of a manageable size. Size reduction is followed by screening, which separates the shredder product into 3 size classes: fine, medium and coarse. The fine section predominately consists of coating and liberated glue and can simply be discarded. The medium size fractions

contain a mixture of coating, glue and PS and are treated with density separation. Coarse material predominately consists of PS particles with little coating. The coating that is present in the coarse size fractions is unliberated, and is subject to selective milling in a hammer mill. The hammer mill product, as well as the density separation product report to secondary shredding, where the remaining glue and coating are liberated and finally separated in the secondary density separation stage. Each unit operation will be discussed in detail. Note that the following discussion does not include secondary shredding, and secondary air separation as the research in these areas is incomplete.

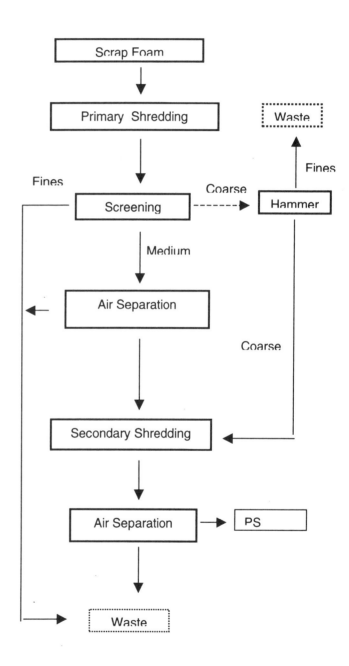

Figure 2 - Flow sheet of process Technology

SHREDDING – Primary shredding serves the function of reducing the casting pattern down to a manageable particulate size. Using shredder screen aperture diameters ranging from 1 to ¼ inch, coating liberation in excess of 95% was obtained. Figure 3 shows percent coating liberation as a function of the shredder exit screen size (top size of the particle distribution). It is apparent that coating liberation increases as the particle top size decreases, which is not very surprising. However, this increase is not dramatic; less than a 2% increase from the 1-inch shred to the ¼-inch shredder product.

the ratio in terms of shred size. Accordingly, the ¾ inch shred displays the highest size segregation, and allows for a maximum amount of coating to be discarded with a minimum of PS. Table 2 shows some values for each of the particle distributions produced from shredding. Notice that for the 1" and ¾" shredded material the particle size that passes 5% of the polystyrene also passes 86% of the coating. On the other hand the ½" and ¼" shredded material passes only 73% and 72% of the coating, which is 13% to 14% less than that obtained with the 1" and ¾" shredded material. Therefore, shredding with a ¾" or 1" exit screen provides a distinct advantage; disposal of fine size fractions will contain a minimal amount of polystyrene, maximizing recovery, while removing a majority of the coating.

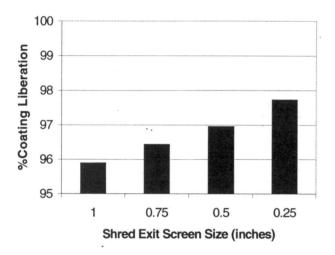

Figure 3 -%Coating liberation as function of shredder screen aperture size.

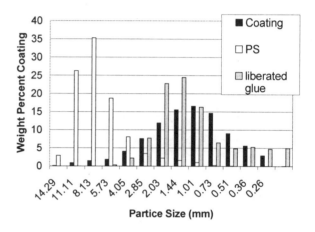

Figure 4 - Size distribution of the ¾ inch shred in terms of weight percent. The black, gray, and hatched bars represent the coating, PS, and liberated glue constituents respectively.

Table 1 shows the mechanical properties of the coating and PS under impact stressing conditions. The expanded PS tends to be resilient due to the cellular structure. Coating, on the other hand, fractures in a brittle mode. Though cutting the patterns with a shearing action, shredding also impacts the material with the blades. This action combined with the mechanical properties of the constituents is the probable cause for such high coating liberation. This dual action of the cutting blades is also manifested in the size distribution of the particles. Figure 4 shows the weight percent of each constituent in each size fraction from the ¾-inch shred. The PS segregates to coarse sizes, while the coating and liberated glue tend to segregate to fine sizes, which is characteristic of brittle materials. The distributions of the constituents follow a similar progression in all other shred sizes as well.

SIZE SEGREGATION – The size distribution of each constituent in Figure 4 suggests that an initial separation of coating and glue from PS can be obtained by simply screening out the fine size fractions. It turns out that size segregation of the constituents is a function of the particle size distribution, which can be controlled with the size of the exit screen on the shredder. Size segregation is quantified by calculating the ratio of d_{50} values of the coating to the polystyrene. Figure 5 shows

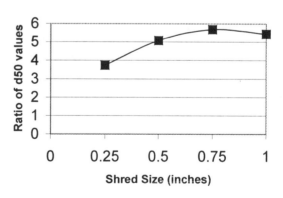

Figure 5 - Ratio of coating d_{50} to PS d_{50} values for the 1, ¾, ½, and ¼ inch shreds.

Table 2. Some characteristic values of the percent coating passing with 5% and 10% passing polystyrene for the 1", ¾", ½", and ¼" shreds.

Aperture diameter of exit screen (inches)	5%	10%	Particle Size (mm)	
			5%	10%
1	86	92	3.4	5.0
¾	86	93	3.1	4.3
½	73	85	1.8	2.9
¼	72	83	1.7	2.2

HAMMER MILLING – A hammer mill, as described above, is generally used in the mining and aggregate industries for crushing soft and friable materials such as limestone. In the case of rejected casting patterns, the hammer mill is used to selectively mill unliberated coating from coarse PS particles. As listed in Table 1, PS particles tend to display elastic or viscoelastic deformation under impact stressing. Coating, on the other hand, exhibits brittle fracture. Due to size segregation, coarse size fractions predominately contain PS particles with unliberated coating and glue. Very little liberated coating is present. It is of interest to subject the coarse size fractions to impact stressing in a hammer mill in order to selectively mill the coating from the PS particles. A separation is then achieved by screening out the fines, which contain the milled coating.

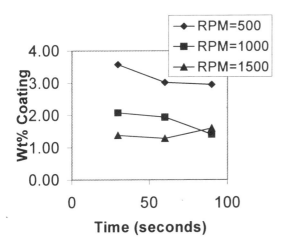

Figure 6 - Weight percent of coating as a function of mill residence time.

Figure 6 shows the weight percent of coating as a function of mill residence time for a mill of RPM 500, 1000, and 1500. The three lines show that mill RPM has an appreciable effect on the removal of coating. However, there appears to be no significant difference between 1000 and 1500 RPM. This suggests that a moderate mill RPM around 1000 RPM will remove a significant amount of coating (75%), while limiting energy input to the mill.

The residence times do not have a very large effect on coating removal over the ranges tested (30-90 seconds). At 500 and 1000 RPM only 0.5wt% more coating is lost as the residence time is increased from 30 to 90 seconds. The 1500 RPM curve actually shows a slight increase. Since increased residence times do not have a significant impact on the coating removal, a minimal residence time is superior as the throughput of material through the mill can be maximized.

DENSITY BASED SEPARATION – The flow sheet in Figure 2 shows fine material reporting to waste, coarse material reporting to hammer milling, and medium material to air separation. Air separation is really the crux of the process as the PS losses in screening and hammer milling are easily controlled. The test results presented here consider a full size distribution rather than a medium size fraction. Though separation of an uncontrolled size distribution in a gravity-settling chamber is not ideal (that is why the flow sheet shows sized material reporting to the air separation unit), it does adequately show the high efficiency of separation in a gravity-settling chamber.

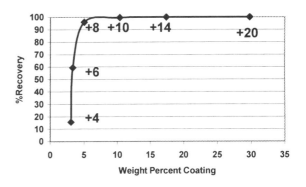

Figure 7 - Cumulative percent recovery of expanded polystyrene plotted against weight percent coating for the third bin in the gravity settling box.

A separation of coating and glue from PS based on density can be performed in either a liquid or air medium. A liquid base medium, such as water, represents an ideal separation of coating from PS as the coating and PS have higher and lower densities than water respectively. Middlings (particles that have coating still adhering to PS) generally float due to the high buoyancy forces. Therefore PS recovery approaches 100%, while separation efficiency also approaches 100%. Overall grade, then, is essentially a function of liberation, as all liberated coating will sink. However, it is undesirable to use a liquid float/sink process. Any material subjected to a float/sink process will require an extra drying step prior to subsequent processing steps, or shipping. Furthermore, emersion of the material in liquid generates another waste stream: particulate contaminated liquid.

The drastic density differences between PS and the contaminants allows for separations in air to be performed with results comparable to float/sink methods without the additional steps required for wet processing. The gravity-settling box deposits material in bins depending upon the particle shape, size, and density. Dense and coarse material settles out in the first few bins, while light and fine material settle out in the latter few bins. PS and coating overlapped in many of the bins due to the wide size distributions, however a relatively clean separation was made by subsequent screening. Figure 7 shows the %recovery of PS in relation to wt% coating retained for the third bin. Each point on the plot represents the percentage of PS in that particular bin retained on the indicated mesh number and the wt%

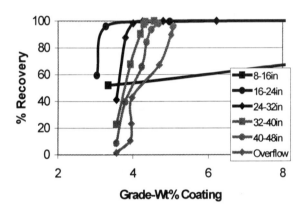

Figure 8 - Cumulative percent recovery vs. wt% coating for all bins

coating. Notice that the line is nearly vertical from +4 mesh to +8 mesh, where in excess of 95% of the PS is retained. This is because almost no liberated coating is present in these size fractions. The coating that is present is essentially unliberated coating. Retaining material below 8-mesh has the opposite effect; little increase in PS recovery is observed, while wt%coating dramatically increases. Figure 8 shows plots for each of the 7 bins in the gravity-settling box, which display similar trends as that of bin 3 shown in Figure 7. Overall 96% of the polystyrene was retained, while containing only 5-wt% coating. Total separation efficiency is estimated as 99%, where approximately 99% of the liberated coating was separated.

CONCLUSION

The combination of shredding, screening, hammer milling, and air separation have provided excellent results. A total of 96% of the polystyrene has been recovered using this technology, while 98% of the coating has been removed. This indicates that a successful approach has been selected for the development of a process technology. Additionally, the high separation efficiencies observed with air separation suggest that the removal of the contaminants is largely a function of the degree of liberation. Furthermore, it has

been observed that shredding with a ¾-inch aperture diameter screen maximizes size segregation. Consequently, a maximum amount of coating can be discarded with a minimum of PS. Hammer milling increases the liberation of coating by selectively breaking unliberated coating from coarse PS particles.

ACKNOWLEDGEMENT

The authors extend their appreciation to the people at GM Powertrain in Saginaw Michigan who supplied financial support as well as samples for testing. Also acknowledged is the USEPA and CenCITT environmental center, which also provided financial support. The authors would finally like to thank Tim Eisele, S. Komar Kawatra, Ron Nuttall, Steve Forsell, Dennis Moore, Brett Krause, and Mitch Loomis of Michigan Technological University and Peter Larson and Martin Toth of the Institute of Materials Processing for their help during the course of the project.

Although the research described in this article has been funded wholly or in part by the U.S. Environmental Protection Agency (USEPA) and the National Center for Clean Industrial and Treatment Technologies (CenCITT), it has not been subjected to USEPA's required peer and policy review and therefore does not necessarily reflect the views of USEPA or CenCITT and no official endorsement should be inferred.

CONTACT

Jeremy Pletka and Jaroslaw Drelich can be reached through the Department of Materials Science and Engineering, Houghton, MI at (906)487-2029 and (906)-487-2932 respectively or email jmpletka@mtu.edu, jwdrelic@mtu.edu.

REFERENCES

1. J. H. Hunter, "Survey Indicates Bull Market for Lost Foam Foundries", Modern Casting, September 1998, pp. 50-52

2. M.J. Lessiter, "Lots of Activity Taking Place Among Lost Foam Job Shops," Modern Casting, April 1997, pp. 28-31.

3. M.J. Lessiter, "Today's lost Foam Technology Differs From Yesteryear," Modern Casting, April 1997, pp. 32-35.

4. N. Margolis, J. Eisenhauer, S. McQueen, and R. Scheer, Report of the "Metalcasting Industry Technology Roadmap Workshop," June 16-17, 1997, Rosemont, IL, Energetics, Inc., Columbia, Maryland, July 1997.

5. B. A. Wills, "Mineral Processing Technology", 1992, 5th edition.

GREENING THE SUPPLY CHAIN

Bottom-Line Advantage Through Management System Integration

Rainer Ochsenkuehn
First Environment, Inc.

ABSTRACT

Integration of management systems allows suppliers to benefit from synergy effects in their efforts to cope with the ISO 14001 certification requirements for their environmental management systems (EMSs). Integration of new management systems, i.e. ISO 14001 into an already existing system, such as ISO 9001/QS-9000, is the most common scenario. This paper outlines methods that follow this common scenario and could be used by small- and medium-sized companies for the integration of management systems.

INTRODUCTION

Ford, GM, Toyota and other OEMs are increasing the pressure on their suppliers to get more environmentally friendly. Greening the supply chain does not address simple product requirements anymore, which specify banned materials and production methods, this concept goes well beyond a strategic influence of the automotive suppliers into their management systems. The ISO 14001 Environmental Management System certification requirements will not only affect the Tier 1 suppliers, but also Tier 2 and below. The challenge is to avoid any costly implementation efforts, but achieve the most effective solution, thereby adding value to the bottom line of the organization. The integration of management systems is a possible opportunity to achieve these goals by taking advantage of synergy effects.

Theoretical models of management system integration have been developed since the early 90's. These models went beyond the typical integration of related areas like environment and health & safety and included quality and other parts of existing management systems. However, these concepts were not taken seriously in the real world of business until recently. A general reason for this current acceptance is the increasing popularity of voluntary international management system standards like ISO 9000, QS-9000 and ISO 14001 which provide an overall framework for management system integration efforts.

These standards have many parallels and similar requirements. The integration of management systems based on these standards and/or other specifications such as OHSAs 18001 and Responsible Care® are not just a paper exercise. Furthermore, existing internal management systems including finance, production or personnel management may be integrated. The effectiveness of the implementation and integration will determine wether organizations can gain efficiency improvements and leverage their experience.

Integrated management systems are now recognized as a significant, value-adding component of the business process. It is clear that many companies are actively pursuing these programs to add value to business processes and to enhance the status of the facilities involved. Now it is recognized that a business value can be derived through the implementation of a sound integrated management system by streamlining processes throughout the organization. The profitability of supplier operationis will will depend on how effective existing management systems are integrated and how flexible the system is in response to future issues, i.e. the addition of a health & safety management system or cost accounting system.

Integration of new management systems in an already existing system is the most common scenario. The following generic concepts in management systems integration will be discussed:

1. 20+0 element integration;

2. Partial integration of management system manuals and procedures;

3. Process driven integration;

4. Modular Management Systems.

20+0 ELEMENT INTEGRATION

At the beginning of the integration movement, organizations simply added the additional ISO 14001 requirements to the existing ISO 9001 elements without any additions or modifications. This seemed to be the easiest way, and at the same time organizations could benefit from the existing experience and efforts in the quality area. This approach of adding all environmental requirements into an additional element, i.e. a 21st chapter, did not prove to be not practical. One of the reasons was the difficulty in accounting for the different types of stakeholders addressed by the different management systems.

PARTIAL INTEGRATION OF MANAGEMENT SYSTEM MANUALS AND PROCEDURES

Another possibility is the method of partial integration. The basis is an existing management system manual and procedures, e.g. a quality management system (QMS). The new elements which are identical or similar requirements will be combined with the existing ones. Elements not fitting in the existing structure will be added as a 21st, 22nd, etc. chapter in the management system manual. With partial integration more effort is necessary at the procedural level. The management system procedures and work instructions need to be modified to ensure the easy understanding and usability of the documents covering all the integrated systems. Figure 1 shows identical system elements between ISO 14001 and ISO 9001.

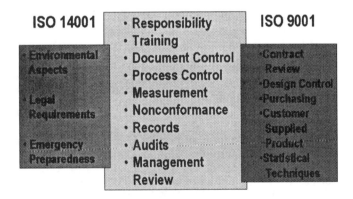

Figure 1. Identical system elements.

PROCESS DRIVEN INTEGRATION

Many organizations have transformed their internal structures through reengineering and lean management to reflect a more process driven orientation. Based on the process model shown in ISO 9000:2000, this integration approach takes an existing process and links the related management system requirements as they are used in the process (see Figure 2). The first level of the process structure divides between the management-, resource-, value-added-, customer- and supporting processes within an organization. The second level of the process structure gives a more detailed description based on specific organizational processes and customer requirements. For example, the management processes include quality and environmental policy, whereas the value-added processes include process control on the quality side and operational control on the environmental side. The responsibility for the management systems requirements has therefore moved to the different process teams and is included in the day-to-day operations.

Figure 2. Process integration opportunities.

MODULAR MANAGEMENT SYSTEMS

The modular based integration approach divides the management systems into groups with identical functions, namely the management functions, production processes and cross-functional elements. Policy, document control and auditing are management and cross-functional examples of encompassing elements which are integrated at the first level. The second level represents the production processes with the specific management system elements. Advantages of the modular integration are higher efficiency, an identical system structure and methodology, and the reduction of complexity. This integration structure is used mostly on a corporate level. It can be mirrored on the facility level as well; however, a process driven integration is also possible.

CONCLUSION

Many of the larger auto suppliers have already begun their ISO 14001 implementation efforts with some plants already certified and recognizing the benefits and potential competitive advantage. Currently small- and mid-sized suppliers are indicating interest, however most do not seem to view the mandate as an imminent issue yet. Such thinking can generate a negative impact as the EMS or integrated system may not be well-thought through and insufficient time may be given for its implementation. The desired benefits ought to be defined up-front and effective measures implemented in order to monitor the management system for continual improvement. An ISO 14001 EMS can only bring the desired results, if the management system concept and the continual improvement philosophy live actively in the organization. The desired benefits will not be realized if the system is implemented soley to achieve certification in order to satisfy customer demands.

Integration Approach	Advantage	Disadvantage
20 + 0 element integration	• Simple integration; • Effortless	• No flexibility; • Not stakeholder specific
Partial integration	• Use of existing structure; • "Unlimited" expansion	• Relatively low flexibility; • Potenital high complexity
Process integration	• Flexible system; • Process thinking; • Easy link with ISO 9000:2000; • Responsibility on "shop flor" level	• Complex integration; • Clear interface definition between processes necessary
Modular integration	• Higher efficiency; • Reduction of complexity	• Clear separation of elements into different groups; • Less practicable on facility level

CONTACT

Rainer Ochsenkuehn is the Strategic Environmental Program Manager for First Environment, Inc. and can be reached at (973)616-9700 x218, or via email: roc@firstenvironment.com.

REFERENCES

1. R. Ochsenkuehn, J. D. Heeren, "Integration, an answer for the new ISO 14001 mandates for Automotive Suppliers?", *Business Standards*, January/February 2000.

2001-01-0623

The Influence of an Oil Recycler on Emissions with Oil Age for a Refuse Truck Using in Service Testing

Gordon E. Andrews and Hu Li
Department of Fuel and Energy, The University of Leeds

J.Hall, A. A. Rahman and P. Mawson
Top High UK Ltd.

ABSTRACT

A method of cleaning lubricating oil on line was investigated using a fine bypass particulate filter followed by an infra red heater. Two bypass filter sizes of 6 and 1 micron were investigated, both filter sizes were effective but the one micron filter had the greatest benefit. This was tested on two nominally identical EURO 1 emissions compliance refuse trucks, fitted with Perkins Phazer 210Ti 6 litre turbocharged intercooled engines and coded as RT320 and RT321. These vehicles had emissions characteristics that were significantly different, in spite of their similar age and total mileage. RT321 showed an apparent heavier black smoke than RT320. Comparison was made with the emissions on the same vehicles and engines with and without the on-line bypass oil recycler. Engine exhaust emissions were measured about every 400 miles. Both vehicles started the test with an oil drain and fresh lubricating oil. The two refuse trucks were tested in a different sequence, the RT320 without the recycler fitted and then fitted later and the RT321 with the recycler fitted and then removed later in the test and both without any oil change. The RT320 was also the one with the finer bypass filter. The test mileage was nearly 8,000 miles both trucks. The air/fuel ratio was worked out by the exhaust gas analysis. The correlation between air/fuel ratio and emission parameters was determined and appropriate corrections were made in the case of that the air/fuel ratio had an effect on emissions. The results showed that the on line oil recycler cleaning system can reduce the rate of increase of the NOx with oil age. There appeared little influence of the oil recycler on carbon monoxide and hydrocarbon emissions. The rate of increase in particulate emissions was reduced by 50% for RT320 and an immediate decrease in particulate emissions was seen on RT320 test after fitting the recycler. The black smoke was reduced by 30% for RT320 in terms average value and an immediate decrease in smoke after fitting the recycler on RT320 test and an immediate increase in smoke after the removal of the recycler on RT321 test were shown.

INTRODUCTION

Lubricating oil forms a significant fraction of the particulate volatile fraction and can contribute to the carbon emissions. Lubricating oil also acts as a sink for carbon emissions (1,2) and unburned diesel fuel and this can lead to the deterioration of the oil (3), which results in an increase in the particulate emissions (1). For a low particulate emissions engine the work of Cooke (4) showed that lubricating oil may contribute more to the carbon emissions than to the solvent fraction at some engine conditions. His results showed that there was a variable influence of lubricating oil with no influence on particulate emissions for some engine conditions and up to a 250% increase with lube oil age for other engine conditions.

Diesel engines with low particulate emissions have a very low lubricating oil consumption. There is a concern that carbon particles may accumulate to a greater mass concentration in the lube oil, as there will be a reduced dilution with top up of the oil (5). High carbon in the lube oil may then increase the contribution of the oil to the particulate emissions through the associated higher viscosity. Andrews et al (6) have shown that a Euro 1 passenger car diesel engine accumulated carbon in the oil at a greater proportion of the carbon emissions than for an older high carbon emitting engine.

The control of combustion chamber deposits (CCD) in modern diesel engines is recognised as a part of low emission engine design. The extended service requirements are making the control of CCD increasingly difficult (7). The primary source of piston deposits is the lubricant (8) and oxidation of the lubricant is the primary cause of deposit formation (9). In cylinder deposits consist of ash from the lube oil additives, carbon and absorbed unburned fuel and lubricating oil (8, 10). The CCD can be a source of wear in engines, increased friction and hence increased fuel consumption. Crownland heavy carbon has been shown to increase oil consumption (11) and deposits have been shown to increase as piston temperature increases above 250C (12). Deposits also increase with the soot content of the oil (12) and hence deposits can increase as the oil ages. The increased soot in oil as it ages results in an increased oil viscosity (13) and this increases the oil (14) and fuel consumption. The aim of the present work was to examine these influences for two Euro 1 refuse trucks fitted with Perkins engines and to determine the improvement in oil quality through the use of on line recycling of the oil to remove soot, wear metals, fuel and water dilution (3). The recycler had a fine bypass oil filter and an infrared heater to distil out water and light fuel fractions. The use of by pass filters is common in some large diesel engines, but is not usual in smaller engines of 6 litres or less.

Andrews et al (1, 6) showed that the lubricating oil age could have a significant influence on particulate emissions. Three IDI engines were tested over 100 hours to investigate the influence of lubricating oil age on the emissions. Two Ford engines, 1.6 and 1.8 litre, were low emission engines and the Petter AA1 engine was an older technology high emissions engine. For all three engines there was little influence of lubricating oil age on gaseous emissions. There was evidence in the NOx emissions for the Petter and Ford 1.8 litre engines of an action of deposit removal, which reduced the NOx and deposit, build up that increased the NOx. This was also supported by the lube oil additive metal analysis. The hydrocarbon emissions increased with oil age for both of the low emission engines but only the 1.6 litre Ford engine showed a similar change in the particulate VOF. The 1.8 litre engine VOF trends were dominated by lube oil influences, which do not contribute to gaseous hydrocarbon emissions at a 180C sample temperature. The particulate emissions trends with oil age were quite different for the Ford 1.6 and 1.8 litre engines, with a continuous increase in emissions for the former and a decrease followed by an increase after 50 hours for the latter. The Petter engine also followed similar trends to the Ford 1.8 litre engine, although with much higher emission levels. It was shown that these trends were also reflected in the carbon fraction and unburned fuel fractions of the particulate VOF for the two Ford engines. However, the lube oil fraction decreased substantially over the first 50 hours for the Ford 1.8 litre engine and then remained at a stable level. The implication was that the fresh lubricating oil resulted in high unburned lube oil particulate VOF emissions and also generated carbon emissions. Once the volatile fraction of the lube oil had been burnt away in the engine, the lube oil VOF remained stable and the subsequent increase in the

particulate emissions was due to increasing carbon emissions. The initial decrease in the fuel VOF fraction followed by an increase after 50 hours was possible due to the initial removal of CCD by the fresh lubricating oil followed by a build up of fresh deposits as the oil aged. Fuel fraction VOF can be contributed to by deposit absorption and desorption, which is a function of the extent of the CCD.

These works is concerned with a technique to keep oil clean, extend its life and reduce the increase in emission that occurs with aged oil. The above review has emphasized the importance of CCDs in emissions and lubricant quality. At the same time as diesel emissions regulations have come into force the trends in the diesel design, towards higher ring zone temperature and pressure, piston redesign for higher top rings, higher piston temperatures, and extended service interval requirements, are making it increasingly difficult to control deposit formation in the engine with traditional oil additive technology (7). Diesel deposits can be classified into two types: ring zone and piston skirt deposits (varnish). The higher ring zone temperatures (325-360C) promote thermal degradation of the lubricating oil and unburned/oxidized fuel components producing a 'carbon' deposit. At relatively low piston skirt temperatures (200-260C) a varnish type deposit predominates (7). The primary source of piston deposits is the lubricant and lubricant oxidation is the primary cause of deposit formation (7). Diesel engine deposits also increase with the oil consumption (8), the piston temperatures (12) and the oil soot content (12). Engine deposits are a source of unburned hydrocarbons through absorption and they also act as a cylinder insulation, which increases NOx emissions because of the higher cylinder temperature (16).

THE ON LINE OIL RECYCLER AND REFUSE TRUCK TEST PROCEDURE

A method of continually cleaning the engine oil on line was investigated. This was based on the combined effects of a bypass fine oil particulate filter with a 1 or 6 micron filter element followed by an infra-red dome heater which heated the oil to 135C as it flowed over a conical cascade into a drain return to the oil sump. The previous work using this system (3) used a 6 micron bypass oil filter and this was the filter used in RT321 at the start of the present work. However, work was in progress to develop a finer I micron bypass filter and this had reached the prototype stage when it was decided to fit a recycler to the second refuse truck (RT320). It had originally been intended in the present work to use two nominally identical refuse trucks with the same engine and mileage. However, as will be shown in the results, the two refuse trucks had different oil deterioration rates and the accompanying emissions results showed different emissions. RT321 was operated consistently 2 units richer in air/fuel ratio than RT320 for the same duty cycle and journey with an average air/fuel of 33/1. Soot accumulation in the oil and lube oil consumption for RT321 were also significantly higher. Consequently, it was concluded that RT321 was in a worse mechanical state to RT320 and the two refuse trucks could not be compared one with a recycler and one without. Thus, both refuse trucks had to be tested with and without the recycler.

RT320 was first tested without the recycler fitted and after 4,700 miles a recycler with the 1 micron bypass filter was fitted without any oil change. RT321 was first fitted with a recycler with the 6 micron bypass filter and after 5,000 miles the recycler was removed without any oil change.

The refuse trucks were operated 6 days a week with an average mileage per day of about 30 miles. However, the engines of refuse trucks were kept in high power output during their work for loading/unloading the rubbish and the refuse trucks had a frequently stop-start duty cycle. These made the engines of refuse trucks work under a severe condition and accelerated the engine oil deterioration, which resulted in increases of engine CCD and thus increase emissions.

One of the advantages of the recycler is that it provides for an improved oil quality and reduced oil consumption (3). The improved oil quality reduces the engine CCDs. Thus the reduced oil comsumption and CCD in the combustion chamber reduce engine emissions. Authors have shown that emissions had been reduced for a Ford1.8 litre IDI engine test as a result of improved oil quality and reduced oil consumption(15).

Other investigators of bypass filters have advocated the improvement by using filters of the order of 1 micron (22-26). The initial choice in the present work of 6 micron particle size filters was based on advice from hydraulic oil filter manufacturers that oil additives could be filtered out if the filter size was too fine. Also, very fine 1 micron oil filters that were available from hydraulic oil filter manufacturers had a rather high pressure loss. If these were used in the recycler then the bypass flow would have been controlled by the filter pressure loss and the flow rate would have decreased as soot built up in the filter. A key feature of the present recycler is that the bypass oil system oil flow rate is relatively high. In the Ford 1.8 litre IDI passenger car diesel tests (3) the recycler oil flow rate was such that oil the sump volume of the oil was passed through the recycler four times an hour. The one micron by pass filter used in this work was developed to have a fine filtration of 1 micron particles without affecting the recycler bypass flow rate, even when loaded with soot. The present results show that this new fine filter does improve the performance of the recycler. This filter is now in production and will be used

in the future commercial use and evaluation work using this system.

Bypass filters have two basic features: a high filtration efficiency and a high particulate storage capacity (24). They reduce the fine particulate matter in the oil leaving the main filter to remove the larger particulates. Stenhouer (25) showed that bypass filters removed organic material, sludge, varnish, resin, soot and unburned fuel. His work showed that over 80% of the contaminant removed by the bypass filter was organic. Bypass filters can remove the small pro-wear contaminant particles thus extending engine component life (26). The benefit of bypass filtration was the reduced wear and reduced in cylinder deposits. It is possible to arrange a combination filter whereby in one housing a main and bypass filter are arranged (24) with the flow between the two splitting according to their flow resistances.

Although a filter based bypass filter was used in the present work, centrifugal bypass filters are also quite common (26). These are of two types: powered and self-powered and the latter are more common in automotive applications. A self powered centrifugal oil cleaner uses the dirty oil pressure to drive the cleaning rotor using centrifugal separation of the high density particles from the lower density oil. These contaminants collect as a hard cake on the inside of the rotor which can then either be cleaned off or disposed of as a unit (26).

Centrifugal filters have an effective particle size removal below 1 micron and have a filtration efficiency that does not deteriorate with time (26). They are generally more expensive initially than filter based bypass filters. One assessment of bypass filters (27) has determined the average size rating (50% removal) of a centrifugal filter as 6-10 microns and hydraulic oil filters as 2 micron. They also estimated that it would take 30 bypass filter changes before the cost of bypass filtration exceeded that of a centrifugal filter. This could be 10 years of normal use. Hydraulic quality bypass filters were used in the present work with an average size rating of 6 and 1 microns.

EXPERIMENTAL TECHNIQUES

ENGINE SPECIFICATIONS - The two refuse trucks tested had the same engine specification that is detailed in Table 1 and operated with routine oil top ups and normal commercial duty cycles. The oil and exhaust gas samplings were taken every two weeks, at an interval of about 400 miles. The same route and driver were used for each test run. However, traffic conditions were varied.

RT320 started the test without the recycler fitted and RT321 started the test with the recycler fitted. For RT320 after 4,700 miles after the commencement of the test with fresh oil, a recycler was fitted. Then the test continued for 3,000 miles. For RT321 after ~5,000 miles after the commencement of the test with fresh oil, the recycler was removed and the test continued for 2,600 miles without the oil being changed. As discussed above, RT320 was fitted with a recycler with a bypass particulate filter size of 1 microns and RT321 was fitted with a 6 micron filter.

FUEL AND LUBRICATING OIL - The fuel used in the tests was commercially available standard low sulphur diesel with sulphur content ≤ 0.05%.

The lubricating oils used in tests were 15W-40 CE/SF mineral oil. The specifications of diesel fuels and lubricating oils are reported in a separate SAE paper.

PARTICULATE SAMPLING SYSTEM - A stainless steel tube (gas sampling tube) with a diameter of 7 mm was inserted into the centre of the tailpipe before the muffler (about one meter from the exhaust manifold). This tube was connected to a relay tube (about 2 meters long stainless steel tube) which was terminated beneath the driver cab and bolted with a stopper when the vehicle was in service. In the case of sampling, a 2 meter rubber tube was used to connect relay tube and the sampling kit which was placed in the cab. Thus the exhaust gases were conducted into the sampling kit. After the completion of the sampling the rubber tube was taken off and the relay tube was blanketed so as not to interfere with the refuse trucks in service.

Table 1. Specifications of the engines

	Refuse truck
Type	Perkins Phazer 210Ti
Maximum Power Rating	156.5KW(210BHP) AT 2500rpm
Displacement litre	6.0
Oil pressure, low idle(min)	
at2100rpm(max)	43-49 P.S.I
Oil capacity litre	18
Bore mm	100
Stroke mm	127
Cylinder No.	6
Compression ratio	17.5:1
Lube oil change intervals	3 months
EGR	
Aspiration	Turbocharge intercooled

The exhaust gases from the rubber tube were conducted through a 125ml wash bottle placed in an ice bucket to condense the water and heavy hydrocarbons and then passed to a filter block with a 47 mm Whatman Glassfiber filter paper to collect the particulate samples. The residue gases were then passed through a gas meter to count the volume of gases and a flow meter, and finally collected in a gas bag. This sampling process was driven by exhaust pressure. Therefore the samples collected were proportional to combustion pressure and represented real driving conditions.

GAS ANALYSIS - The exhaust gases collected in a 60 litre gas sample bag were analysed as soon as possible after sampling. The gases were passed to an oven and then transported to a heated FID at 180°C for total hydrocarbon analysis, a heated Chemiluminescence NOx analyser for NOx analysis, a Servomex paramagnetic analyser for oxygen analysis and a Hartman & Broun Uras 10E for CO and CO_2 analysis. The air/fuel ratio could be worked out by these gas analyses according to carbon balance principle.

PARTICULATE ANALYSIS - The particulate filter papers were conditioned in a constant humidity enclosure for 24 hours before and after the tests and the weights recorded after each 24 hour conditioning period. The increase in weight was the particulate mass and this was measured to 1 microgram accuracy and a minimum mass of 1 mg was collected, giving a minimum resolution of the mass of 0.2%. Only one filter paper was taken at each oil mileage as the driver and vehicle were only available for one test journey.

The particulate was analysed for carbon, fuel and unburnt lubricating oil fraction using TGA. A round cutter with a diameter of 29 mm was used to cut the filter papers so as to get an identical area of the filter paper and minimise the interference of blank filter papers. This cut filter paper sample was wired and hung on a hook on an end of a microbalance enclosed in the oven with nitrogen atmosphere. The sample was heated up to 550°C at a rate of 20°C per minute and kept for 10 minutes where no further weight loss occurred. The weight loss represented the volatile fraction of particulates. The air was then introduced and the temperature was increased to 560°C and maintained for 20 minutes. The weight loss before and after introduction of the air was equivalent to the carbon mass. The rest of the sample was the ash mass of particulates. A blank filter paper was used to determine the weight loss of the filter paper and this was used to correct the particulate weight loss. The TGA procedure has been used for determination of fuel and lubricating oil fractions of particulates and detailed in reference 15.

BLACK SMOKE MEASUREMENT - An OPAX 2000-II smoke meter was used in bus and refuse truck tests. This is a partial flow smokemeter or opacimeter designed for measuring the exhaust smoke of diesel engines. The principal configuration is that it has a heated smoke measuring chamber which contains a light source (halogen lamp), a sensor (silicon photodiode with corrected spectral response similar to the photopic curve of the human eye - peak response 550nm) and a mirror as shown in Fig.3.5. It measures the smoke(opacity) based on the visible light absorption of the exhaust gas as detected by a sensor.

The testing procedure used was the Free Acceleration Smoke (FAS) measurement required by the Vehicle Inspectorate for MOT test. This procedure has good repeatability and considered to indicate the vehicle's actual smoke emission. The details are as follows:

Smoke testing procedure:

-Switch OPAX 2000-II smoke meter on to warm it up.

-At the end of the warm up period the OPAX 2000-II shows dashes on displays.

-Press the MEASURE key. The display shows "REF" and the equipment starts a new autozero. After this, the equipment is ready to operate.

-Start the engine which should already be at running temperature. Depress the accelerator pedal until the engine attains its governed speed and then release the accelerator to idle. Repeat this several times in order to purge the exhaust of loose dust and carbon.

-Insert the probe in the exhaust pipe using the adapter provided. Ensure it is fixed to the tailpipe.

-Press the MEASURE key. When the display is flashing "-1-" the operator can commence the accelerations.

-For each acceleration, the operator depresses the accelerator sharply to full fuel position and on attaining governed speed remains there for approximately 2 sec, then releases the accelerator. The display will delay for 5-6 sec and then prompts the operator for a new acceleration; blinking on the MEASURE display indicates the next acceleration is required.

-Repeat the acceleration and idling procedure for a minimum of six accelerations. An average value of opacity is displayed. Print it out and the test is finished.

-The unit measured by this procedure was K (1/m) from 0.00 to 9.99.

AIR/FUEL RATIO

As the operation of engines could not be the same for every sampling cycle, even for the same route and driver, the air/fuel ratio had to be corrected to the mean air fuel ratio for each test journey. The air fuel ratio was determined by carbon balance from the exhaust gas analysis.

Fig.1 shows the air/fuel ratio for two truck tests. The variation of air/fuel ratio was from 32 to 38 with an average value of 34.2 for RT320 and from 29 to 35 with an average value of 32.3 for RT321. This shows that RT321 had a 2-unit richer combustion condition than RT320. The average air/fuel ratio for two refuse truck tests is 33.2, which was used as a mean air/fuel ratio to correct raw emission results if a good correlation between emission parameter and air/fuel ratio exists.

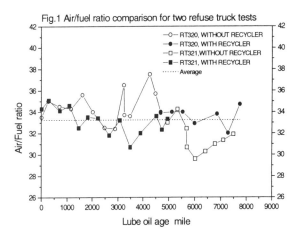

Fig.1 Air/fuel ratio comparison for two refuse truck tests

The gaseous, smoke and particulate emissions were calculated against the air/fuel ratio using linear regression analysis to determine the correlation between them. The regressed results are shown in Figs.2 and 3. It has shown that the following emission parameters have a significant correlation with the air/fuel ratio:

For RT320:   NOx;

For RT321: CO, HC, total particulate mass, particulate ash, particulate carbon, particulate VOF and lube oil VOF.

It revealed that most emission parameters for RT320 were not sensitive to air/fuel ratio whereas in contrast, most emission parameters were correlated to air/fuel ratio for RT321.

GASEOUS EMISSIONS RESULTS

CARBON MONOXIDE - As shown in Figs.2a and 3a, the air/fuel ratio did not have a significant effect on CO emissions for RT320 and yet a notable effect for RT321. Therefore, the CO emissions were corrected to the mean air/fuel ratio for RT321.

The CO emissions for two refuse trucks were shown in Figs.4 and 5. For RT320, without the recycler the CO emissions declined slightly from 240 ppm to 220 ppm with two peaks appearing at around 1,100 miles and 4,500 miles respectively. With the recycler fitted, the CO emissions continued to decrease from 220 ppm to 190 ppm after 1,300 miles using the recycler. Then two consecutive peaks in CO emissions appeared followed by falling back to 190 ppm. The first peak appeared at 1,100 miles was associated with a fall in NOx emissions and a peak in hydrocarbon emissions shown in Figs.7.48 and 50, which could be explained as due to low temperature in combustion chamber. The other peaks were not mirrored in NOx and hydrocarbon emissions and could be due to traffic variations. For RT321, without the recycler the CO emissions fluctuated with an increasing trend (from 360 to 390 ppm in 5,000 miles) with oil age. With the recycler removed, the raw data were showing a higher CO emissions. However, as CO emissions were affected by the air/fuel ratio, the data was corrected to the mean air/fuel ratio and showed that CO emissions were at a similar level. Comparing two refuse trucks, it could be found that refuse truck RT321 had higher CO emissions (about 150 ppm higher) than that for RT320. This indicated that the combustion process in the RT321 engine was more incomplete.

HYDROCARBONS - The total gaseous hydrocarbon emissions were correlated to the air/fuel ratio only for RT321, as shown in Figs.2b and 3b. Thus the raw data for RT320 and air/fuel ratio corrected total hydrocarbon emissions for RT321 were shown in Figs.6 and 7.

The hydrocarbon emissions decreased with oil age generally for RT320 without the recycler. There was a peak at about 1,100 miles of oil age. This peak in HC emissions was corresponding to a peak in CO emissions and a fall in NOx emissions at the same time. This indicated that the temperature in the combustion chamber at this point was low and thus gave rise to a low NOx but high CO emissions and HC emissions. This is a typical example of emission variation due to traffic variation. With the recycler fitted, the total gaseous hydrocarbon emissions continued to decrease until 6,000 miles of oil age, followed by an increase after 6,900 miles, which was co-ordinated with an increase in CO emissions at the same time. The reason for this increase was not very clear, possibly due to large amount of oil top up and traffic variation. Generally, the fitting of the oil

recycler did not alter the gaseous hydrocarbon emissions for RT320.

The total gaseous hydrocarbon emissions from RT321 test with the recycler fitted showed a generally constant level with some fluctuations. After the recycler was taken off, hydrocarbon emissions had a large increase after 800 miles, where the hydrocarbon emissions were almost doubled, followed by a rapid fall. Another large peak in hydrocarbon emissions appeared after 1,300 miles. This refuse truck had a very high smoke emissions, as shown below, and a high lube oil consumption. Hence it must have some mechanical faults in its engine. These two peaks in hydrocarbon emissions could be due to excessive unburnt lube oil.

NOx EMISSIONS - The NOx emissions were sensitive to the air/fuel ratio for RT320 and yet not sensitive to the air/fuel ratio for RT321 as shown in Figs.3c and 3c. Hence NOx emissions were corrected to the mean air/fuel ratio for RT320. Figs.8 and 9 show the raw and corrected NOx emissions for two refuse trucks.

For RT320 without the recycler the NOx emissions increased with oil age in general. There was a decrease at around 1,100 miles of oil age, which was associated with a peak in CO and HC emissions that could be due to incomplete combustion in cylinder. Afterwards there was a peak for NOx emissions followed by a decrease. The general trend was illustrated by a linear trendline and the gradient was 15 ppm/kilomile. After the recycler was fitted, the NOx emissions were stabilised and the increase of NOx with oil age was 5 ppm/kilomile. The reduction in the rate of increase of NOx emissions by the recycler was 67%.

For RT321, with the recycler the NOx emissions showed a fluctuated variation with oil age. There was a large peak at around 2,500-3,000 miles of oil age. However, it can be found that a similar peak appeared on the RT320 test at a similar oil age. This similarity indicated that this peak could be caused by deposit accumulation. The peak value, however, lasted longer without the recycler on the RT320 than that with the recycler on the RT321 and the decline in NOx after the peak was larger on the RT321 with the recycler, which suggested that the recycler had removed more deposits from the cylinder and thus led to a more significant decrease in NOx emissions.

The NOx emissions after the recycler had been taken off from the RT321 were showing an increasing trend, although there were some variations. The rate of increase in NOx was 3 ppm/kilomile with the recycler and 22 ppm/kilomile without the recycler. However, the data with the recycler was quite scattered.

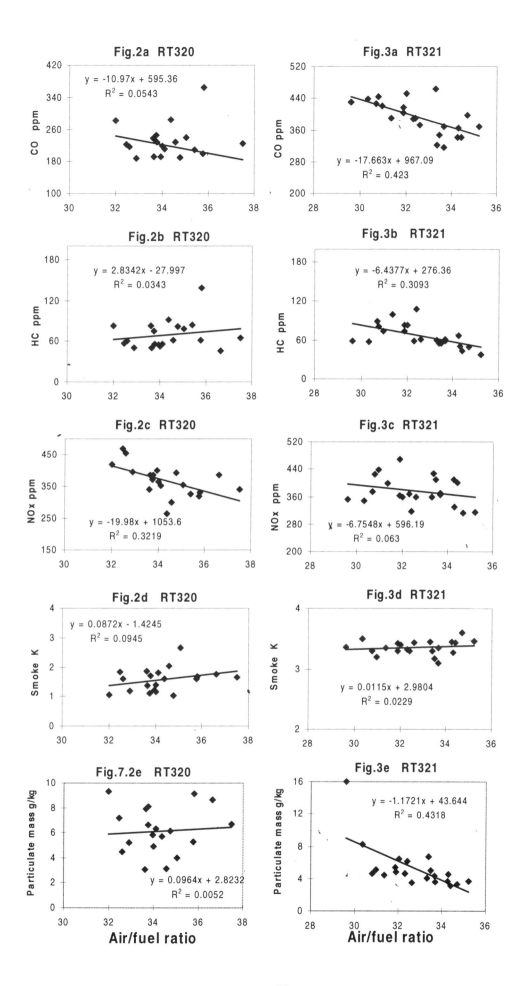

Fig.2a RT320
Fig.3a RT321
Fig.2b RT320
Fig.3b RT321
Fig.2c RT320
Fig.3c RT321
Fig.2d RT320
Fig.3d RT321
Fig.7.2e RT320
Fig.3e RT321

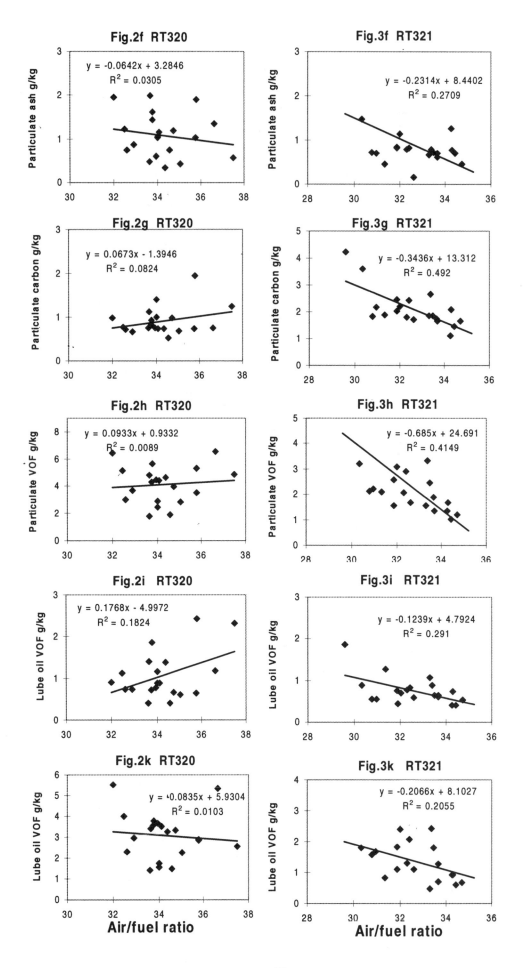

Fig.2f RT320
Particulate ash g/kg
y = -0.0642x + 3.2846
R² = 0.0305

Fig.3f RT321
Particulate ash g/kg
y = -0.2314x + 8.4402
R² = 0.2709

Fig.2g RT320
Particulate carbon g/kg
y = 0.0673x - 1.3946
R² = 0.0824

Fig.3g RT321
Particulate carbon g/kg
y = -0.3436x + 13.312
R² = 0.492

Fig.2h RT320
Particulate VOF g/kg
y = 0.0933x + 0.9332
R² = 0.0089

Fig.3h RT321
Particulate VOF g/kg
y = -0.685x + 24.691
R² = 0.4149

Fig.2i RT320
Lube oil VOF g/kg
y = 0.1768x - 4.9972
R² = 0.1824

Fig.3i RT321
Lube oil VOF g/kg
y = -0.1239x + 4.7924
R² = 0.291

Fig.2k RT320
Lube oil VOF g/kg
y = -0.0835x + 5.9304
R² = 0.0103

Fig.3k RT321
Lube oil VOF g/kg
y = -0.2066x + 8.1027
R² = 0.2055

Air/fuel ratio

Air/fuel ratio

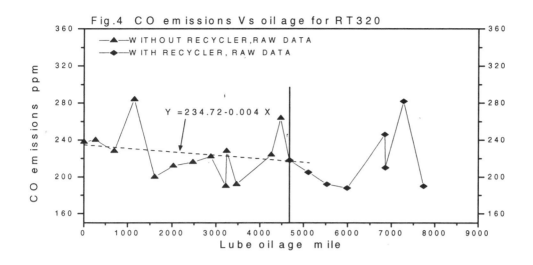

Fig.4 CO emissions Vs oil age for RT320

WITHOUT RECYCLER,RAW DATA
WITH RECYCLER, RAW DATA

Y =234.72-0.004 X

CO emissions ppm

Lube oil age mile

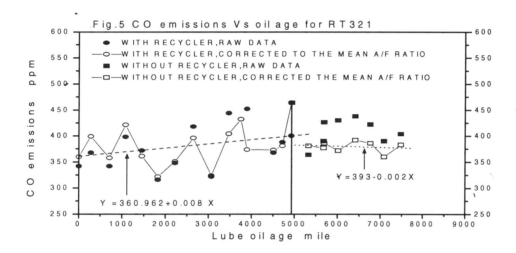

Fig.5 CO emissions Vs oil age for RT321

WITH RECYCLER,RAW DATA
WITH RECYCLER,CORRECTED TO THE MEAN A/F RATIO
WITHOUT RECYCLER,RAW DATA
WITHOUT RECYCLER,CORRECTED THE MEAN A/F RATIO

Y =360.962+0.008 X

Y =393-0.002X

CO emissions ppm

Lube oil age mile

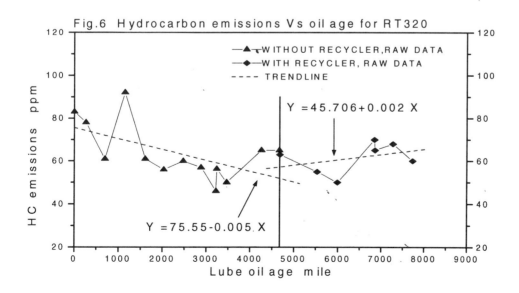

Fig.6 Hydrocarbon emissions Vs oil age for RT320

WITHOUT RECYCLER,RAW DATA
WITH RECYCLER, RAW DATA
TRENDLINE

Y =45.706+0.002 X

Y =75.55-0.005 X

HC emissions ppm

Lube oil age mile

BLACK SMOKE

Black smokes were not sensitive to the air/fuel ratio within the range of tests as shown in Figs.2d and 3d. The R^2 was 0.09 and 0.02 for two refuse trucks respectively. So the raw data was used to show smoke variation with oil age.

Figs.10 and 11 show the black smoke varied as a function of oil age for two refuse trucks. For RT320 without the recycler, a sharp peak appeared at 270 miles of oil age, which was attributed to the traffic condition. Except this data, the smoke fluctuated around an average value of 1.7 K until 3,500 miles of oil age, where the smoke started to increase continuously. The average value for smoke during 4,700 miles of testing without the recycler for RT320 was 1.7 K. After the recycler was fitted, the smoke was sharply reduced from 1.9 K down to 1.4 K, 8 miles of running with the recycler fitted. Afterwards the smoke showed a gradually declined trend with oil age. The average level of black smoke was 1.2 K with the recycler fitted. Hence a reduction of 0.5 K was achieved by the recycler, which was a 30% reduction.

Refuse truck RT321 showed very high black smoke, twice as high as that on the RT320. The use of the oil recycler had shown an effect on reducing the smoke gradually with oil age. The smoke was about 3.5 k for fresh oil and declined to 3.2 k after nearly 5,000 miles of oil age. After the recycler had been taken off the smoke increased from 3.2 k to 3.5 k just after 8 miles of travel. Then the smoke decreased and increased periodically around an average value of 3.4 K. Thus the reduction in smoke by the recycler could be worked out by comparing the smoke just before and after taking off of the recycler. This was 0.3 K.

In sum, the recycler has shown an apparent improvement on the reduction of black smoke. The reduction on smoke for RT320 was 30% and 9% for RT321. The less improvement for RT321 was due to very high smoke for baseline test. With the recycler fitted the smoke decreased with oil age for both trucks whereas the baseline tests for both trucks showed a constant level with some fluctuations. The recycler reduced the smoke very rapidly after its fitting and the smoke deteriorated in a very short period when the recycler was taken off.

PARTICULATE EMISSIONS

TOTAL PARTICULATE EMISSIONS - It has been shown in Figs.2e and 3e that the total particulate emissions were sensitive to the air/fuel ratio only for RT321. Therefore the raw data was plotted against the oil age for RT320 as shown in Fig.12 The raw and air/fuel ratio corrected total particulate mass was plotted against the oil age for RT321 as shown in Fig.13.

For RT320, without the recycler the total particulate emissions increased with the oil age with a series of increases and decreases. The mass of total particulates increased from 3 g/kgfuel of fresh oil to 9 g/kgfuel after 4,700 miles of oil age. The average rate of increase in total particulate emissions was 1 g/kgfuel per kilomile. After the recycler was fitted to RT320, the particulate mass was reduced by 46% immediately followed by slow increases with oil age. There was a peak at about 7,300 miles of oil age or 2,600 miles after the recycler was fitted, where the total particulate mass was increased by about 3 g/kg or 35%. This peak was associated with an increase in particulate ash and VOF discussed later. During the testing period with the recycler fitted, the rate of increase in total particulate mass was 0.5 g/kg per kilomile on average, excluding the peak point. Thus the reduction of increase rate in particulate mass by the recycler was 0.5 g/kg per kilomile or 50% for RT320.

The total particulate emissions were showing an increasing trend with a series of increases and decreases with oil ageing on the RT321 test with the recycler. The average rate of increase in particulate mass was 0.4 g/kg per kilomile. After the recycler was taken off, there was a huge peak at 6,000 miles of oil age, where the mass of particulates reached 16 g/kg and was about three times higher than average level. It is noted that this peak in particulate mass was following the peak in hydrocarbons (appeared at 5,700 miles). This could be explained that certain amount of hydrocarbons from lube oil stored in the cylinder due to incomplete combustion could gave rise to a peak in hydrocarbon emissions. Then these lube oils were burned and resulted in an increase in total particulate emissions. It will be shown later that the particulate ash, carbon and VOF emissions increased at the same time, which indicated the excessive lube oil was burned. As the excessive lube oil was burnt, more ash would be produced and some went into particulates and some would went into the oil. Thus there would be an increase in additive metals in the oil. The oil analysis proved this when the calcium and zinc in oil reached a peak.

PARTICULATE ASH EMISSIONS - There was no good correlation between particulate ash emissions and air/fuel ratio. Therefore the raw data was used to show the particulate ash emissions as a function of oil age in Figs.14 and 15.

Without the recycler RT320 showed an apparently increasing trend in particulate ash emissions with oil age. The rate of increase in particulate ash mass was 0.2 g/kg per kilomile. After fitting the oil recycler the mass of the particulate ash decreased very quickly and remarkably. The fall in the mass of particulate ash before and after the fitting of the recycler was from 1.91 g/kg down to 1.06 g/kg. i.e. 45% of instant reduction by the recycler.

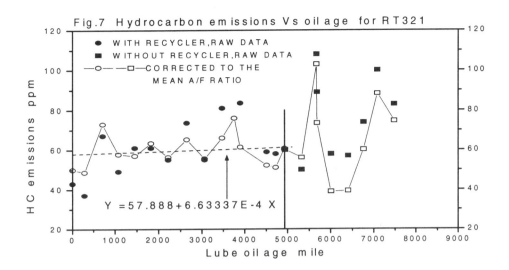

Fig.7 Hydrocarbon emissions Vs oil age for RT321

$Y = 57.888 + 6.63337E-4\ X$

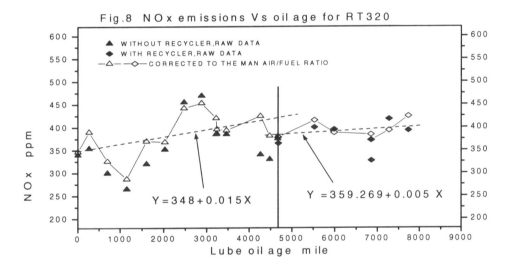

Fig.8 NOx emissions Vs oil age for RT320

$Y = 348 + 0.015 X$

$Y = 359.269 + 0.005\ X$

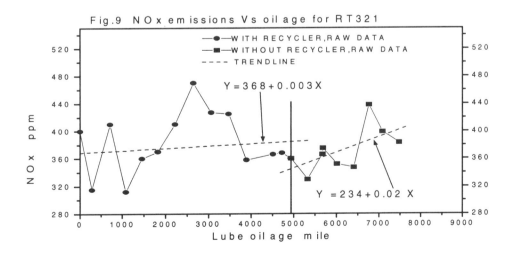

Fig.9 NOx emissions Vs oil age for RT321

$Y = 368 + 0.003 X$

$Y = 234 + 0.02\ X$

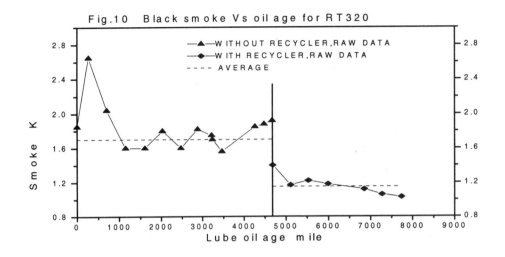

Fig.10   Black smoke Vs oil age for RT320

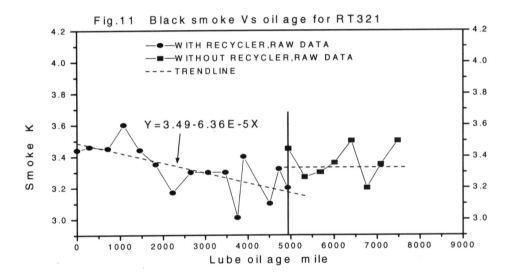

Fig.11   Black smoke Vs oil age for RT321

$Y = 3.49 - 6.36E - 5X$

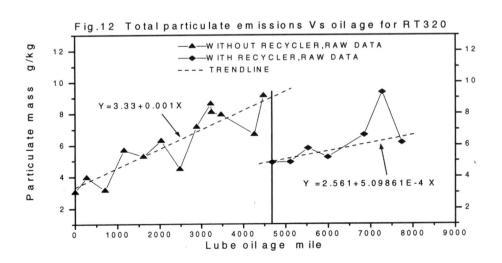

Fig.12   Total particulate emissions Vs oil age for RT320

$Y = 3.33 + 0.001X$

$Y = 2.561 + 5.09861E - 4 X$

This reduction in particulate ash mass suggested that the recycler reduced the lube oil consumption. The particulate ash then had a decrease after 800 miles with the recycler fitted and from here started to increase continuously until it reached a peak at 7,300 miles. This peak was associated with the increase in total particulate emissions. The general trend in particulate ash with the recycler was a slightly increasing trend.

For RT321, with the recycler the particulate ash was firstly increased and then decreased in the first 1,900 miles of oil age. Afterwards it was stabilised except for a small peak at 4,000 miles. The average level for particulate ash emissions was 0.75 g/kg with the recycler. After taking off of the recycler, the particulate ash had a large peak at 6,000 miles, where the mass of particulate ash was over four times higher than average level and this peak was associated with the peak in total particulate emissions. Except this peak, there was no significant difference in particulate ash emissions before and after the fitting of the oil recycler on RT321 test.

PARTICULATE CARBON EMISSIONS - There was no correlation between air/fuel ratio and particulate carbon emissions for RT320 and yet the correlation existed for RT321 as shown in Figs.2g and 3g. The raw and air/fuel ratio corrected data was demonstrated in Figs.16 and 17.

For RT320 without the recycler fitted, the particulate carbon had an initial decrease in the first 700 miles and then stabilised in the next 2000 miles. There was a remarkable increase in particulate carbons in the last 1,500 miles, which was corresponding to the increasing trend in black smoke. After the recycler was fitted, the mass of particulate carbon decreased immediately and continued to decrease for another 1,300 miles, followed by a small increase to 1.0 g/kg and then stabilised.

For RT321 with the recycler, particulate carbon emissions were showing a constant trend fluctuated around a mean value of 1.75 g/kg. After the recycler was taken off, the particulate carbon started to increase and reached a peak with a doubled value at the oil age of 6,000 miles, followed by a fall back to the average level. This huge peak was associated with the total particulate and particulate ash peaks.

The level of particulate carbon emissions for RT321 was doubled, compared to the results for RT320. This was co-ordinated with the black smoke emissions discussed earlier, which showed the smoke for RT321 was twice as high as for RT320.

PARTICULATE VOF EMISSIONS - Particulate VOF (Volatile Organic Fractions) was determined by the TGA technique and had a good correlation with the air/fuel ratio for RT321 (R^2=0.41), but no correlation for RT320(R^2=0.01) as shown in Fig.2h and 3h. Thus the raw data of particulate VOF was plotted against oil age for RT320, and the raw and air/fuel ratio corrected data of particulate VOF was plotted against oil age for RT321 as shown in Figs.18 and 19.

For RT320 without the recycler the particulate VOF emissions fluctuated periodically. Its variation was corresponding to that of total particulate mass. The general trend was an increasing tendency. The gradient for increase in particulate VOF was 0.78 g/kg per kilomile. After the fitting of the recycler, the particulate VOF mass was reduced by 55% immediately. there was a significant peak at 7,300 miles of oil age and a small peak at 5,500 miles. These two peaks were corresponding to the total particulate emission peaks at the same time. The general trend in particulate VOF emissions was increasing with the oil ageing and the average rate of increase was 0.77 g/kg per kilomile. Hence the rate of increase in particulate VOF emissions was not changed by the recycler.

For RT321 with the recycler, the particulate VOF mass increased slowly with less variation compared to that for RT320. There was a huge peak at 6,000 miles of oil age where the recycler had been taken off. This peak also appeared in particulate ash and carbon emissions as discussed above. it is postulated that this peak was due to mis-function of the engine.

These two refuse trucks behaved quite differently in particulate VOF emissions. For RT321 the particulate VOF mass was less oil age dependent, compared to that for RT320. The overall particulate VOF emissions for RT321 was much lower than that for RT320.

LUBE OIL VOF EMISSIONS - The particulate lube oil VOF emissions appeared correlated to the air/fuel ratio only for RT321, shown in Figs.2i and 3i. Figs.20 and 21 show the raw and corrected data for two trucks.

For RT320, without the recycler the particulate lube oil VOF emissions showed an increasing trend from 0.5 g/kg of fresh oil to 2.5 g/kg after 4,700 miles. The average rate of increase was 0.35 g/kg per kilomile. There was a peak at 1,100 miles of oil age, where CO, hydrocarbons, smoke were also showing a peak. This indicated that the temperature in combustion chamber was low for some reason so that more incomplete combustion products were produced. After the recycler was fitted, there was an immediate decrease in lube oil VOF emissions and the trend was showing a continuous decrease with oil age. The fall in lube oil VOF by the recycler was 65% by mass.

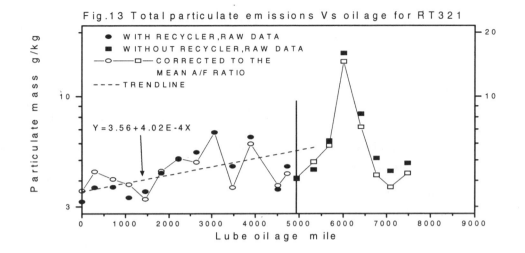

Fig.13 Total particulate emissions Vs oil age for RT321

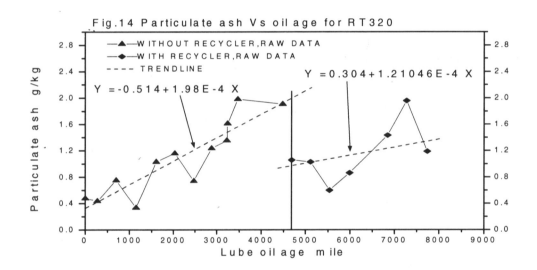

Fig.14 Particulate ash Vs oil age for RT320

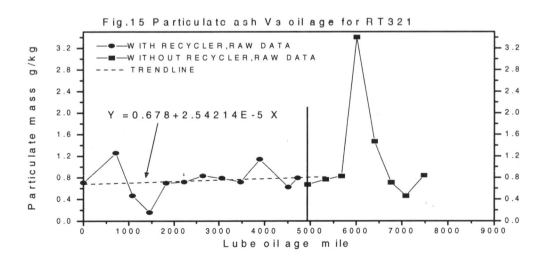

Fig.15 Particulate ash Vs oil age for RT321

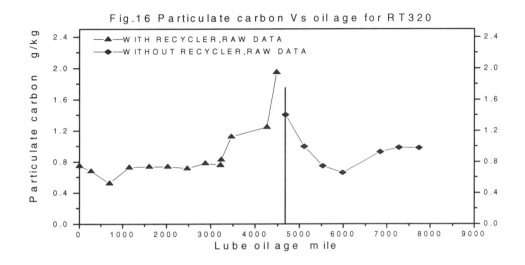

Fig.16 Particulate carbon Vs oil age for RT320

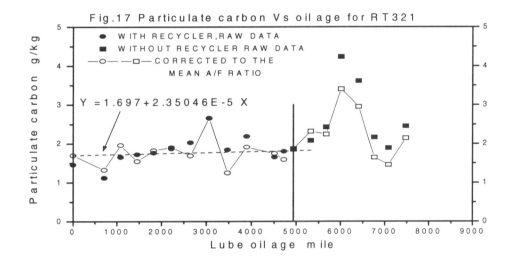

Fig.17 Particulate carbon Vs oil age for RT321

$Y = 1.697 + 2.35046E - 5 \ X$

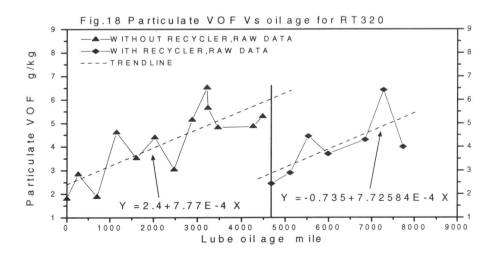

Fig.18 Particulate VOF Vs oil age for RT320

$Y = 2.4 + 7.77E - 4 \ X$

$Y = -0.735 + 7.72584E - 4 \ X$

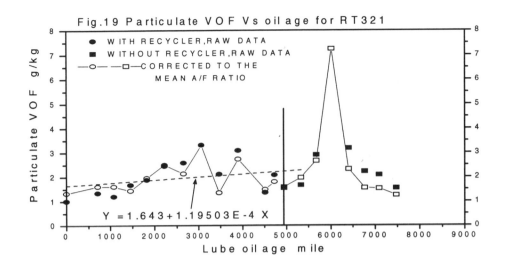

Fig.19 Particulate VOF Vs oil age for RT321

$Y = 1.643 + 1.19503E-4\ X$

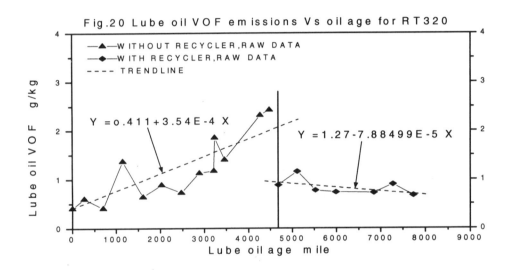

Fig.20 Lube oil VOF emissions Vs oil age for RT320

$Y = o.411 + 3.54E-4\ X$

$Y = 1.27 - 7.88499E-5\ X$

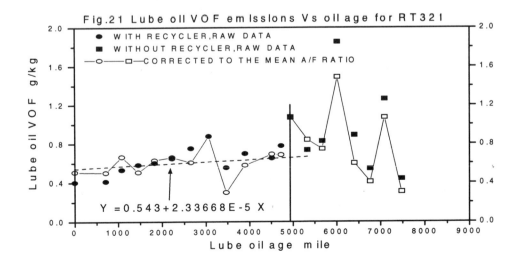

Fig.21 Lube oil VOF emissions Vs oil age for RT321

$Y = 0.543 + 2.33668E-5\ X$

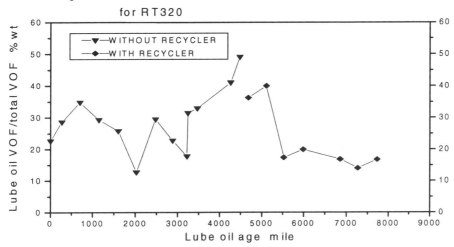

Fig.22 Fraction of lube oil VOF to total VOF emissions for RT320

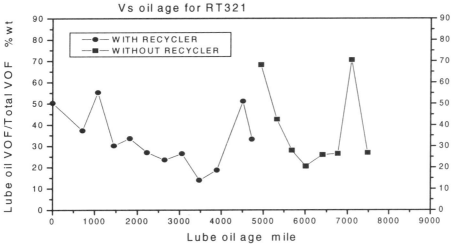

Fig.23 Fraction of lube oil VOF to total VOF emissions Vs oil age for RT321

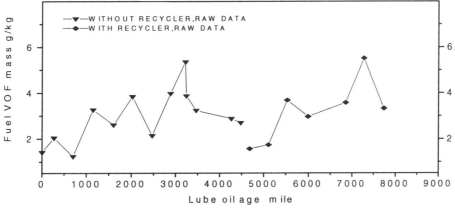

Fig.24 Diesel fuel VOF Vs oil age for RT320

For RT321, with the recycler fitted the lube oil VOF was showing a slightly increasing trend around 0.6 g/kg. After the recycler was taken off, there was an immediate increase (43%) in lube oil VOF mass emissions followed by decreasing trend. A large peak appeared at 6,000 miles, which was associated with the peaks in particulate ash and carbon emissions. Differing from other emission parameters, there was a second peak at 7,000 miles, which was only appeared for lube oil VOF. This indicated there were lube oil remains in the combustion chamber, possibly due to absorption of lube oil on deposits. The deposits were increased after the taking off of the recycler, which was proved by the NOx emission rise in Fig.7.49. Thus more lube oils could be absorbed and then released.

Figs.22 and 23 illustrated the lube oil VOF as a percentage of total particulate VOF for two refuse trucks. For RT320 without the recycler the lube oil VOF took up about 20-35% of total VOF. The fraction of lube oil VOF had a 10% drop after the recycler was fitted and further decreased to 15 to 20% after 800 miles of operation of the recycler. For RT321, with the recycler the fraction of lube oil VOF decreased continuously from fresh oil till 4,000 miles. With the recycler taken off, the fraction of lube oil VOF immediately increased to 70% and then decreased. A peak appeared at 7,000 miles of oil age where the lube oil VOF rose up to 70%.

UNBURNT FUEL VOF EMISSIONS - There was no good correlation between fuel VOF emissions and air/fuel ration as shown in Fig.2k and 3k. Hence the raw data was presented as a function of oil age.

The diesel VOF showed an increasing trend with oil age for RT320 tests both with and without the recycler. Similarly, there was an immediate decrease in diesel VOF mass after the fitting of the recycler. The diesel VOF was relatively stable for RT321 tests, except a huge peak appeared at the oil age of 6,00 miles. The diesel VOF mass was higher for RT320 than that for RT321. The details were shown in Figs.24 and 25.

COMPOSITION OF PARTICULATE EMISSIONS - Particulate mass emissions were analysed for particulate ash, carbon, total VOF, lube oil VOF and fuel VOF emission as shown above. All these components of particulates were integrated into a single plot to see the composition of the particulate emissions as a function of oil age as shown in Figs.26 and 27.

It has shown that for RT320 the variation of total particulate mass with oil age was primarily depending on the total VOF mass variation and secondarily on the ash mass variation. The variation of particulate carbon mass was independent to total particulate mass. The variation of total VOF mass was mainly decided by the fuel VOF mass variation. Similarly for 321, the variation of total particulate mass with oil age was primarily depending on the total VOF mass variation and yet secondarily depending on both ash and carbon mass variation. The VOF emissions was mainly affected by the fuel VOF variation. A huge peak at 6,000 miles appeared at all phases of particulate emissions.

The composition analysis were showing that for RT320, the particulate VOF took up 60-80% of total particulate emissions. The particulate ash varied from 10-30% and carbon varied from 10-20% of total particulate mass. Among the total VOF fraction, the fuel VOF contributed 40-60% to total particulate mass and lube oil VOF had a contribution of 10-20 % to total particulate mass. The fitting of an oil recycler on RT320 initially reduced total VOF fraction due to a reduction of lube oil VOF fraction but increased ash fraction and at late stage the ash fraction was reduced and the VOF fraction was recovered to the level of without the recycler. The recycler had reduced lube oil contributions to total particulate mass in terms of percentage.

For RT321, a high carbon contribution of 30-50% to total particulate emissions was revealed. The ash fractions were between 10-20%. The total VOF fractions were about 40-50%. Compared to the results from RT320, particulate emissions for RT321 revealed a higher carbon, lower total VOF, lower fuel VOF and similar ash fractions. The high carbon emissions resulted in high black smoke. The taking off of the recycler from the RT321 led to an increase in the fractions of particulate carbon.

PYROLYSIS GAS CHROMTOGRAPHY OF PARTICULATES - The pyro-probe GC was used to analyse the particulate samples to show the carbon number distribution of particulate VOF fractions. Fig.28 shows the profiles of particulates with and without the recycler for RT320. Fig.29 shows the results for RT321 test. It clearly shows that fuel peaks ranging from C13-C26 and lube oil peaks ranging from C20 upwards existed for two refuse trucks. For RT320, without the recycler it shows more lube oil fractions ranging around C27 and slightly less fuel fractions ranging from C13-C15 existed, compared to the profile with the recycler. For RT321, without the recycler the slightly more fuel fractions ranging from C13-C15 and slightly more lube oil fractions appeared, compared to the results with the recycler .

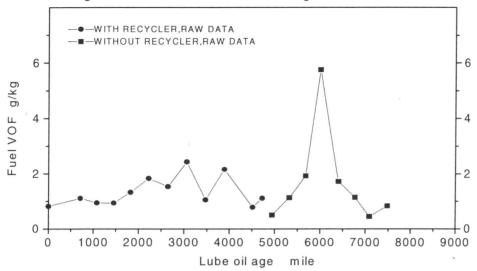

Fig.25 Diesel fuel VOF Vs oil age for RT321

CONCLUSIONS

Two refuse trucks with a similar age were selected and the tests were carried out to evaluate the influence of the TOP-HIGH lube oil recycler on emissions and lube oil quality as well as lube oil and fuel consumption. The two refuse trucks were 1991 model fitted with Perkins Phazer 210Ti series turbocharger intercooled engines with EURO-I emission compliance. These two refuse trucks carried out similar duties and used the same mineral lube oils and low sulphur diesel fuels. The lube oil and emissions samples were taken every two weeks or 400 miles on average and analysed. The results are showing that with the recycler:

There is no improvement on carbon monoxide emissions.

NOx emissions varied with oil age. The recycler reduced the rate of increase in NOx emissions.

No notable effect on gaseous hydrocarbon emissions.

The recycler shows a significant reduction on black smoke, not only the absolute value but the rate of increase. With the fitting of a 1 micron pore size filter recycler, the black smoke decreased by 30%.

The increase rate of total particulate emissions has been reduced by 50%. An immediate 46% decrease in particulate mass has been observed after fitting the recycler.

Particulate ash emissions were reduced by 50%. The rate of increase in particulate ash has been slowed down by 40%.

Particulate carbon emissions were reduced. The RT321 had a doubled particulate carbon emissions and thus doubled black smoke, compared to the results for RT320.

A reduction of 55% in particulate total VOF emissions has been achieved .

The lube oil VOF emissions were reduced by 30-65%.

The RT320 had higher VOF emissions whereas the RT321 had higher carbon emissions and thus higher black smoke emissions.

The 1 micron pore size filter in the recycler has shown a higher efficiency on improving oil quality and consequently reducing the emissions than 6 micron size filter.

ACKNOWLEDGEMENTS

We would like to thank Top High UK for a research scholarship to Hu Li and for a research contract. We would like to thank Manchester City Council, Gorton Depot for their co-operation in this work and in particular to John Stanway, General Manager and Joe Stephens, Works Manager. The technical assistance was given by Ian Langstaff and Geoff Cole.

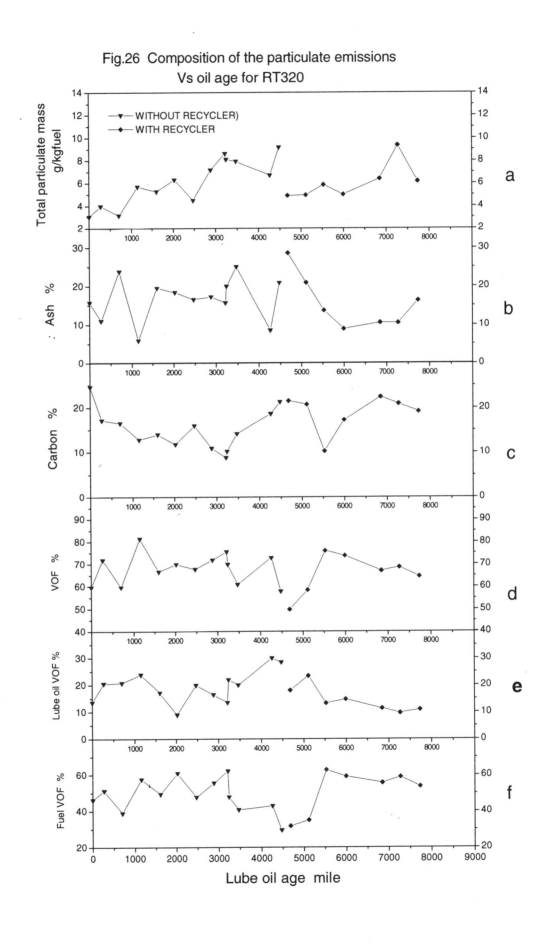

Fig.26 Composition of the particulate emissions Vs oil age for RT320

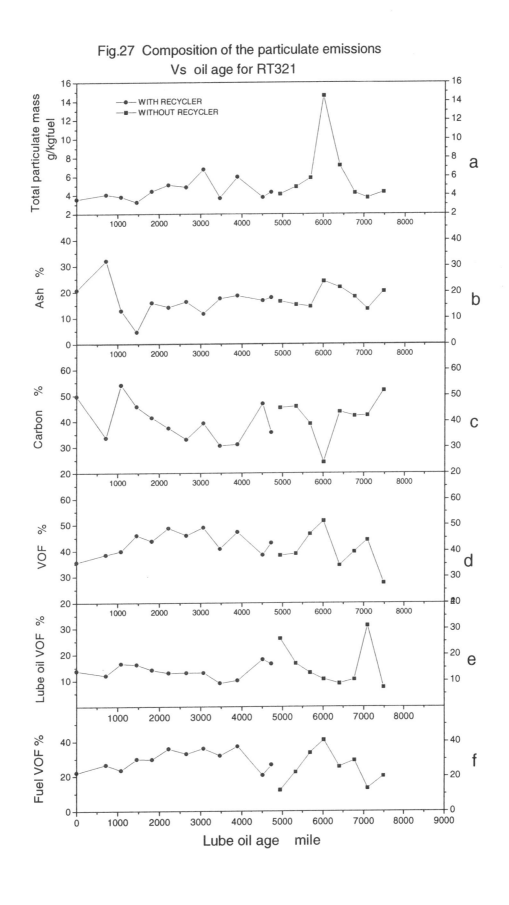

Fig.27 Composition of the particulate emissions
Vs oil age for RT321

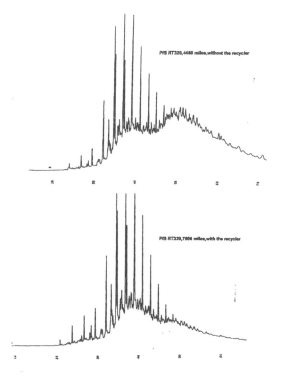

Fig. 28   Pyro-probe GC of particulates for RT 320

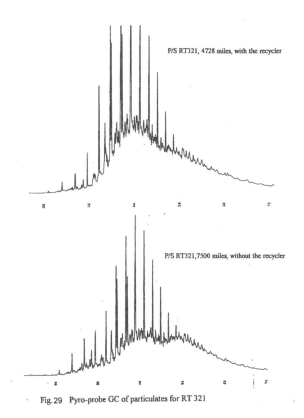

Fig. 29   Pyro-probe GC of particulates for RT 321

REFERENCES

1. Andrews, G.E., Abdelhalim, S. and Williams, P.T., The influence of lubricating oil age on emissions from an IDI diesel, SAE Paper 931003, 1993.
2. Sun, R., Kittleson, D.B. and Blackshear, P.L., Size distribution of diesel soot in the lubricating oil, SAE Paper 912344, 1991.
3. Andrews, G.E., Li Hu, Xu, J., Jones, M., Saykali, S., Abdul Rahman, A. and Hall, J., Oil quality in diesel engines with on line oil cleaning using a heated lubricating oil recycler, SAE Paper 1999-01-1139, 1999
4. Cooke, V.B., Lubrication of low emission diesel engines, SAE Paper 900814, 1990.
5. Cartellier, W. And Tritthart, P., Particulate analysis of light duty diesel engines with particular reference to the lube oil particulate emissions, SAE Paper 840418, 1984.
6. Andrews, G.E., Abdelhalim, S.M. and Hu Li, The influence of lubricating oil age on oil quality and emissions from IDI passenger car diesels, SAE Paper 1999-01-1135,1999.
7. Hutchings, M., Chasan, D., Burke, R., Odorisio, P., Rovani, M. and Wang, W., Heavy duty diesel deposit control – prevention as a cure, SAE 972954, 1997.
8. Graham, J.P. and Evans, B., Effects of intake valve deposits on driveability, SAE Paper 922200, 1992.
9. Andrews, G.E., Abdelhalim, S., Abbass, M.K., Asadi-Aghdam, H.R., Williams, P.T. and Bartle, K.D., The role of exhaust pipe and incylinder deposits on diesel particulate composition, SAE Paper 921648, 1992.
10. Boone, E.F., Petterman, G.P. and Schetelich, Low sulfur fuel with lower ash lubricating oils – A new recipe for heavy duty diesels, SAE 922200, 1992.
11. Hercamp, R.D., Premature loss of oil consumption control in a heavy duty diesel engine, SAE Paper 831720, 1987.
12. Kim, Joong-Soo; Min, Byung-Soon; Lee, Doo-Soon; Oh, Dae-Yoon; Choi, Jae-Kwon, The characteristics of carbon deposit formation in piston top ring groove of gasoline and diesel engine, SAE 980526, 1998.
13. Bardasz, E.A.; Carrick, V.A.; George, H.F.; Graf, M.M.; Kornbrekke, R.E.; Pocinki, S.B.; Understanding soot mediated oil thickening through designed experimentation - Pa–t 4: Mack T-8 test, SAE 971693, 1997.
14. Van Dam, W.; Kleiser, W.M.; Lubricant related factors controlling oil consumption in diesel engines, SAE 952547, 1995.
15. Andrews, G.E., Hu Li, Hall, J., Rahman, A.A. and Saykali, S., The influence of an on line heated lubricating oil recycler on emissions from an IDI passenger car diesel as function of oil age, SAE Paper 2000-01-0232, 2000.

16. Harpster, M.J.; Matas, S.E.; Fry, J.H.; Litzinger, T.A., An experimental study of fuel composition and combustion chamber deposit effects on emissions from a spark ignition engine, SAE Paper 950740, 195.

17. McGeehan, J.A. and Fontana, B.J., Effect of soot on piston deposits and crankcase oils – Infrared spectrometric technique for analysing soot, SAE Paper 901368, 1990.

18. K.Iwakata, Y.Onodera, K.Mihara and S.Ohkawa, "Nitro-oxidation of lubricating oil in heavy-duty diesel engine", SAE 932839

19. Bardasz, Ewa A.; Carrick, Virginia A.; George, Herman F.; Graf, Michelle M.; Kornbrekke, Ralph E.; Pocinki, Sara B., "Understanding soot mediated oil thickening through designed experimentation~Part 5: knowledge enhancement in the General Motors 6.5 L", SAE 972952

20. Bardasz, Ewa A.; Carrick, Virginia A.; George, Herman F.; Graf, Michelle M.; Kornbrekke, Ralph E.; Pocinki, Sara B., "Understanding soot mediated oil thickening through designed experimentation~Part 4: Mack T-8 test", SAE 971693

21. Parker,D.D. and Crooks,C.S., "Crankshaft bearing lubrication formation component manufacturer's perspective", IMechE Seminar on Automotive Lubricants: Recent Advances and Future Developments, S606, 1998.

22. York, M.E., "Extending engine life and reducing maintenance through the use of a mobile oil refiner", SAE831317

23. Byron Lefebvre, "Impact of electric mobile oil refiners on reducing engine and hydraulic equipment wear and eliminating environmentally dangerous waste oil", SAE 942032

24. Loftis, Ted S.; Lanius, Mike B., "A new method for combination full-flow and bypass filtration: venturi combo", SAE 972957

25. Stehouwer,D.M., "Effects of extended service intervals on filters in diesel engines", Proc. International Filtration Conference-The unknown Commodity, Southwest Research Institute, July 1996

26. Samways,A.L. and Cox,I.M., "A method for meaningfully evaluating the performance of a by-pass centrifugal oil cleaner", SAE 980872

27. Verdegan, Barry M.; Schwandt, Brian W.; Holm, Christopher E.; Fallon, Stephen L., "Protecting engines and the environment~A comparison of oil filtration alternatives", SAE 970551

28. Covitch, M.J., Humphrey, B.K. and Ripple, D.E., Oil thickening in the Mack T-7 engine test – fuel effects and the influence of lubricant additives on soot aggregation, SAE Paper 852126, 1985.

29. Ripple, D.E. and Guzauskas, J.F., Fuel sulphur effect on diesel engine lubrication, SAE Paper 902175, 1990.

30. Hartmann, J. High performance automotive fuels and fluids, Motor books International, p. 24, 1996.

31. Caines, A. and Haycock, H., Automotive Lubricants Reference Book, 1996.

CONTACT:

Professor G E Andrew, Dept. of Fuel and Energy, The University of Leeds, Leeds LS7 1ET, U.K.
Tel: 0044 113 2332493 Email: g.e.andrews@leeds.ac.uk

Dr Hu Li, Dept. of Fuel and Energy, The University of Leeds, Leeds LS7 1ET, U.K.
Tel: 0044 113 2440491 Email: fuelh@leeds.ac.uk

Mr James Hall, Top High (UK) Ltd, 918, Yeovil Road, Slough SL1 4JG , U.K
Tel: 0044 01753 573400 Email: tophigh@tophigh.co.uk

2001-01-0625

Real-Time Integrated Economic and Environmental Performance Monitoring of a Production Facility

Bert Bras, Scott Duncan, Matt Franz, Tom Graver, Yong-Hee Han, Leon McGinnis, Sandra Velasquez, Brian Wilgenbusch and Chen Zhou
Georgia Institute of Technology

Dave Gustashaw
Interface Americas, Inc.

ABSTRACT

In this paper, we describe our work and experiences with integrating environmental and economic performance monitoring in a production facility of Interface Flooring Systems, Inc. The objective of the work is to create a 'dashboard' that integrates environmental and economic monitoring and assessment of manufacturing processes, and provides engineers and managers an easy to use tool for obtaining valid, comparable assessment results that can be used to direct attention towards necessary changes. To this purpose, we build upon existing and familiar cost management principles, in particular Activity-Based Costing and Management (ABC&ABM), and we extend those into environmental management in order to obtain a combined economic and environmental performance measurement framework (called Activity-Based Cost and Environmental Management). Furthermore, we enhance the response time of performance measurements by integrating data gathering capabilities from manufacturing process sensors. In this paper, we describe our motivation for this approach, as well as the lessons learned in the implementation.

WHAT IS THE GAS MILEAGE OF YOUR COMPANY?

Public opinion, legislation, and an increasing pressure to become ISO 14000 compliant force manufacturers to include environmental considerations into their decision-making process or risk a variety of financial penalties. Clearly, the recent rise in oil prices has caused several industries to realize their vulnerability to rising energy costs and to become much more interested in assessing their energy consumption. And some companies are realizing that they can achieve a competitive advantage through better environmental performance.

But how well are companies performing? All companies measure economic performance. However, few have even started to measure environmental performance beyond compliance and even fewer know the relationship between environmental and economic performance. A case in point; when asked, most members of the Society of Automotive Engineers can state the gas mileage of their cars. Dave Gustashaw, Vice-President of Engineering for Interface Americas, however, likes to ask his colleagues in industry: "What is the "gas mileage" of your *company*?" Basically, the question is whether they know their company's efficiency in terms of first quality products produced over materials needed. Although many give an answer, most readily confess that it is mostly a guess. It was this question that led us to take the car analogy one step further and think about how we could help decision makers with a tool for answering this question.

BUILDING A DASHBOARD FOR ECONOMIC AND ENVIRONMENTAL PERFORMANCE MONITORING

In order to make environmentally conscious design and manufacturing (ECDM) decisions, companies must be able to a) assess, quantitatively, their full environmental load, b) relate their environmental load to specific design and operations decisions, c) evaluate the economic impact of making different decisions, and d) do all this in a timely fashion. We have partnered with one of the leading industrial proponents of sustainability – Interface, Inc.- in a program of research and development that attempts to address these four issues in a novel way by integrating industrial controls, activity-based costing, process modeling, logistics modeling, and internet technology. With Interface, we are developing a test-bed for integrated environmental and economic monitoring in a real manufacturing plant.

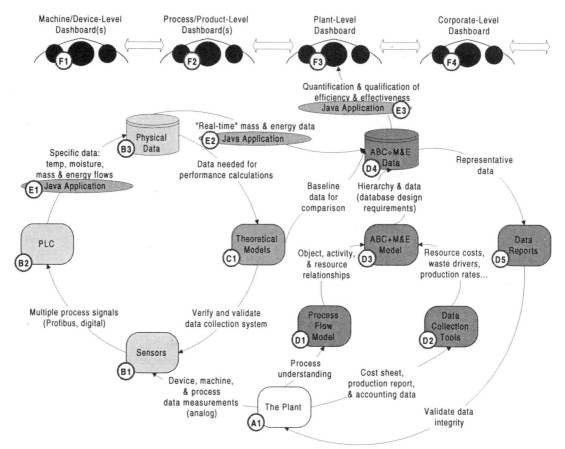

Figure 1 - Integrated Economic and Environmental Assessment Framework

In essence, we are building a "dashboard" similar to a car's dashboard (or instrument panel) that will give a plant manager (and others) with a single glance an idea of how their plant is performing from both an environmental and an economical perspective. Think about your car's dashboard; it gives you basic information and you quickly notice when you are within or outside certain operational limits (e.g., RPM limits). The notion behind our factory dashboard is very similar; we want a tool that quickly displays to a plant manager whether his/her plant is operating within some predefined economic and environmental operational limits, and to gain deeper understanding about the critical economic and environmental parameters in a manufacturing operation, as well as their trade-offs and correlation.

The "dashboard" as we conceive it provides not only a "current" report of the plant operations, but also uses and gives access to historical data. The historical data serve two important functions: first, they provide a basis for establishing "norms" for operating data; second, they provide a basis for trend analysis and comparisons across different time periods, e.g., the same quarter in different years. Ultimately, we expect the dashboard prototype to evolve into a decision support system that will enable users to ask important "what if?" questions, e.g., about alternative materials, alternative processes, even alternative product mixes.

In Figure 1, a schematic of the framework of this integrated test-bed is shown. The main idea is that at the top level, control panels or "dashboards" (labeled F1 through F4) provide decision-makers a read-out of a manufacturing plant's economic and environmental performance analogous to the way an automobile dashboard provides information to a driver about the performance of the car. The idea of a dashboard type performance monitoring has also been mentioned in (Kaplan and Norton, 1996), but no implementation has been made so far. We call our dashboard the EcoDash, where eco stands for both economical and ecological, as well as for its original Greek root, that is, "house". Thus, we are talking about a "house" dashboard, which is quite appropriate because that is what it is all about: we want to know what is happening inside the "house", that is, the physical factory.

Given that there are decision-makers at multiple levels in a company, we anticipate that they will have different information needs, and thus different implementations of dashboards, ranging from the machine/device operator level for machine tuning to the corporate level for strategic decision-making. In order to get the data needed to feed the read-outs of the dashboards, we need to obtain:

a. physical data (labeled B3 in Fig. 1) from the plant (labeled A1) using sensors, PLCs, master controllers, etc., that is, the "regular" factory control systems.

b. financial data from the general ledger, purchasing, production, and other accounting systems.

We build our models based on the Activity-Based Cost and Environmental Management (ABCEM) approach described in [1]. By combining the physical and financial information, we create an Activity-Based Cost, Mass and Energy model (labeled D3) that contains the necessary cost, energy, and mass data (labeled D4) needed for display in the dashboards. Java applications (labeled E1 through E3) handle the communication through a distributed network. A number of theoretical models (labeled C1) are typically required in order to calibrate and validate findings and results.

The intended functionality of the system is:
a. To show near-real-time process data for rates, settings, efficiency, and performance indicators.
b. To allow the user to query an Activity Based Cost and Environmental Management model that is actively populated by both real-time and manually entered data, supporting comparisons of resource consumptions and costs between activities and/or products.

Generally speaking, we are creating a manufacturing process tracking system to meet the information needs of anyone from the low-level manufacturing line operator to the high level CEO. Costing information is merged into the system via the Activity-Based Costing structure. Data from sources including real-time process sensors, production plans, cost sheets, and the accounting general ledger are combined as necessary to support a variety of user queries. However, we are NOT building an Enterprise Resource Planning (ERP) system or another control system. We are focusing on combining information FROM ERP and control systems in a more meaningful and useful way for environmental and economic decision-making. We like to derive the necessary information as much as possible from existing company sources, such as general ledger, ERP, and factory control systems.

DEVELOPMENTAL APPROACH

To achieve the above framework for economic and environmental monitoring, we are proceeding along the following steps:
1. Develop a model of the plant's processes using an Activity-Based Cost, Mass, and Energy approach
2. Develop the mechanisms to obtain physical data from the plant's processes using sensors, master controllers, etc.
3. Integrate the data streams into a dashboard that provides read-outs of the plant's performance at various levels of detail.
4. Develop theoretical models to verify the validity of our models and results.
5. Assess our findings and lessons learned with respect to developing an integrated economic and environmental assessment using this test-bed.

In the following sections, some background information on these steps is given, but we start with describing the production facility that is our living laboratory.

THE PRODUCTION PLANT

Interface, Inc. is one of the world's leading flooring systems manufacturer with 29 manufacturing sites, dispersed globally. Their products are sold in more than 110 countries and in almost every country, they enjoy a market share of roughly 40% in the commercial sector. Interface and its founder, Ray Anderson, are very much committed to becoming more environmentally friendly. Due to the size of the company, we have focused on one manufacturing site located in central Georgia. There are hundreds, if not thousands, of product variants produced. However, the majority are tufted product lines. These are dominating sales and can essentially be described as being manufactured by two-dimensional huge sewing machines. The production volume is measured in square yards [yd^2] and not in physical discrete units like in all other case studies. This is due to the fact that carpet and textile manufacturing is essentially like chemical processes where the products are not discrete units.

MODELING THE PLANT

Our first step was to learn and model what is going on inside the plant. We have used the Activity-Based Cost and Environmental Management (ABCEM) approach to model the facility. This approach is an extension to Activity-Based Costing and includes not only cost, but also energy and waste assessments (see [1,2]. Like Activity-Based Costing (ABC), resources are consumed by activities, and activities are consumed by (cost) objects, see Figure 2. The difference between "classical" ABC and ABCEM is of course that we also perform energy and waste assessments in addition to cost assessments. In general, ABCEM has the advantage that management gets interested almost immediately in the work because the modeling activity will provide them a superior costing model as well, in addition to insight on how energy and mass consumptions correlate with costs.

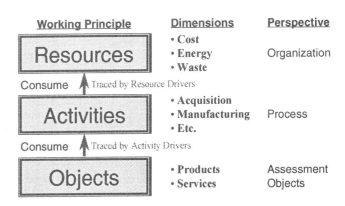

Figure 2 - Working Principle of Activity-Based Cost and Environmental Management [1, 2]

In-plant data sources were used extensively, including the existing standard cost data and production data reporting system. Note that, in general, the ABCEM models would draw from existing data sources, such as an ERP. Thus, we are not in conflict with contemporary trends in business software development, but build upon existing systems.

PROCESS MONITORING AND DATA INTEGRATION

Given that we have an ABCEM model, a key issue is to efficiently obtain the physical data from the processes in the plant that we can use in the model. We have chosen to use web-based technologies. The wide acceptance of browser based interfaces has lowered the learning curve for using this kind of software. However, web-based applications are growing slowly in the manufacturing arena compared to the dramatic blooming in e-commerce. Although there are many ad-hoc web-based applications, there are few systematic studies of the design and architectures of web-applications in manufacturing. In addition to the conservative nature of the manufacturing community, there are significant hurdles in developing applications in manufacturing systems, such as hard real-time interface requirements, complex operation requirements on the server side, proprietary interfaces on the equipment, closed applications on the shop floor and stringent safety and security concerns. The basic web technologies, URL, HTTP and HTML, are not sufficient to support performance monitoring applications. In order to provide efficient data acquisition for different purposes at different levels, we provide two types of data channels (see Figure 3):

1. Direct channels between the sensors and monitor, which are suitable for infrequent but critical data.
2. Relay via a database. The data flow from database to the user is triggered by user's request.

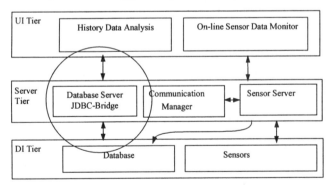

Figure 3 - Process Monitoring Architecture.

Although we use both approaches, the latter is easier to implement because equipment and sensor connections can involve a large variety of proprietary interfaces, and we do not intend to provide many drivers. Thus, we use a database as a data buffer between the monitoring tool and the factory sensors and PLCs. A set of configurable web components provides the linkage between this data base and the monitoring tool. This database is also useful for complex analysis using historical data.

THEORETICAL MODELS FOR VALIDATION AND CALIBRATION

A key issue is to ensure validity of the data and results. For this purposes we develop some theoretical models that will allow us to see whether the data obtained is meaningful and correct. For example, the carpet coating and drying process uses more than half of the total energy required to produce the carpet products. A theoretical model complements the sensors that are installed to monitor energy use, mass inflows, and waste stream contents. The combination of the data collection system and the model provides us with the tools needed to gain understanding into the physics of the process and to validate the information sent to the dashboards.

TOP LEVEL MONITORING TOOL

Given that we have developed the underlying data-acquisition infrastructure and modeling, our next step was to integrate this data into a usable "dashboard" tool that provides the desired factory performance monitoring capability. This task was broken down into the following sub-tasks, some of which are still ongoing.

Develop user interface: This is a web-browser type user interface to the monitoring system. Although many possible implementations are possible, our main goal was to obtain a prototype as fast as possible for proving our concept. It will go online in November 2000. We are leaving user-interface refinements for later.

Identify Metrics and Indicators to display: A key issue to decide is what to display. In addition to absolute energy consumption and waste generation, we would also like to provide indicators useful for benchmarking and trend analyses. A number of different options are prototyped in collaboration with Interface, Inc. Many combinations between mass, dollars, and energy are possible. For assessing environmental impact using a single number, there are a number of different environmental indicators in existence or under development in both the US and Europe (e.g., the Global Warming Potential, the Eco-Indicator, the Environmental Point System, the Global Reporting Index). All have their advantages and disadvantages.

Use the tool and gain understanding into environmental and economic trade-off: Once a prototype "dashboard" has been established, we can start learning about environmental and economic trade-offs and win-win situations. We plan to study the operations at the Interface plant for an extended period of time and record the fluctuations in cost, energy, and mass metrics and indicators. Having the web-based interface allows us to perform remote monitoring. Clearly, Interface is deeply interested in a fundamental understanding of the mass, energy, and dollar flows through their factory, and how strong the correlation is. Furthermore, we plan to study the differences between the production of different types of carpets. Interface has a number of different carpet products with a large variety of color schemes that can be made to order. Standard times and recipes are used, but with the monitoring capability, these could be verified daily.

IMPLEMENTATION DETAILS AND STATUS

Our first prototype will go online in the factory in November 2000. Currently, we are completing the last sensor installations, debugging some communications connections, and implementing the Graphical User Interface (GUI) to database connections. A schematic of the current implementation of the EcoDash is given in Figure 4. Each element is explained next.

A – Sensor Data: For proof of concept, we are monitoring one activity within Interface's facility - the coating and drying line. This activity applies latex and Protekt to the carpet in addition to drying these substances. The parameters are either sent to the physical database (F) or can be read directly by the Dashboard. The sensed mass, energy, and performance parameters include:
- Inflow gas, latex, Protekt, electricity
- Carpet temperature, moisture, velocity
- Exhaust temperature, mass

B – Production Data: This data includes the quantity produced of each object. It is separated by activity, aggregated by day, and entered daily.

C – Accounting Data: Entered once a month, this information is derived from bills and other sources that indicate price fluctuations such as:
- Water, gas and electricity bills
- Raw material costs
- Labor rates

D - Cost Data: Derived from cost sheets, this data includes object specific standards. It may be entered once for each new product, or as product information changes. Data includes:
- Yarn type and amount
- Latex amount
- Labor amount

E – Engineering Data: This information consists of miscellaneous estimations, calculations and formulas that support the ABC model. Specific examples include:
- Latex recipe
- Waste as percentage
- By-products from gas combustion

F – Physical Data: An Oracle database holds the "real-time" data from the sensors. A datum point is entered every second for each sensor. Data is also aggregated by hour and by day to reduce the amount of data.

G – Activity-Based Costing + Mass and Energy Data: An Access database stores information concerning the Activity-Based model. Forms allow the user to input data from B, C, D, and E. It also automatically stores aggregated data once a day from the physical database (F). This database retains historical data to allow the user to view trends.

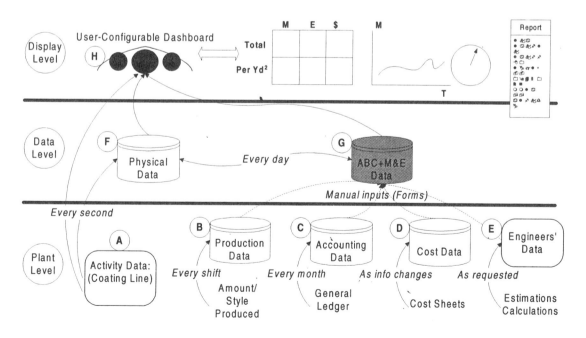

Figure 4 – EcoDash Implementation Schematic

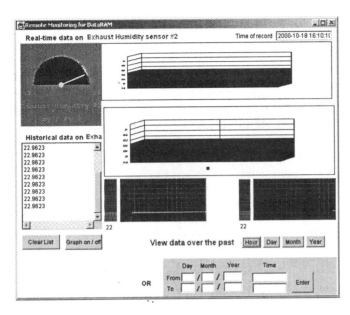

Figure 5 - Screenshot of a Dashboard Window

H – The Dashboard: The dashboard consists of dials, graphs, and tables that display the requested information. Either database (F or G) may supply the requested information. In Figure 5, a sample screenshot is given of a window that is used to display data from the "physical data" database (F). Figure 5 is not a main screen, but a sub-screen that would be accessed when a user wants to see more information regarding the actual sensor data. We are currently completing our GUI for displaying the data from the ABC database and screenshots are forthcoming. When obtaining data from the ABC database, a request consists of the following items:

- Resource consumed (All, specific resource...)
- Objects (All products, specific product)
- Activities (All activities, specific activity)
- Time period (year, month, week, day...)
- View (cost, mass, energy, Eco-Indicator 95...)

The results can be in the form of a table and/or a graph. From the physical database, the request consists of the following:

- Resource consumed (All, specific resource...)
- Activities (All activities, specific activity)
- Time period (month, week, day, real time...)
- A table, graph, and/or dial displays the results.

A report generator can (of course) also be added.

ANTICIPATED/DESIRED USER SCENARIOS

The following user scenarios give a good overview of the desired capabilities of the EcoDash dashboard. It is intended as a decision support tool for many types of users. The Dashboard will either provide the answer or provide insight to help answer the following questions and requests for each type of user.

Worker:
Job: To ensure that activity operates properly
- Is the activity working properly?
- Is the activity producing quality products?
- Is the mass and energy efficiency within acceptable limits?

Activity/Plant Supervisor
Job: Responsible for producing first quality products at all times for one or multiple activities
- Is the activity effective?
  - Is each component working properly?
  - What are the parameters of the activity that affect performance (ie. temp, humidity, speed)? Are they acceptable?
- Is the activity efficient?
  - What is the current mass and energy efficiency? Show this for each resource.
- How does this relate to past trends?

Manufacturing Engineering
Job: Produce products in an economically and environmentally efficient manner
- Are all activities running effectively?
- How does the efficiency compare with historical data?
- Can the efficiency be increased without affecting performance?
- Are there trends in resource consumption?
- How much energy and matter will we consume in the next week? Month? Year?
- How much does each activity cost? What is the breakdown of these costs (i.e., labor, resources, overhead)?
- How much waste are we producing? From what activity? Does the waste vary by product? Why?
- What is the economic and environmental cost of current waste streams- to dispose of and to purchase originally?

Product Design Engineer
Job: To design products, including selection of materials and appropriate processes
- Currently, what is the environmental impact of each of our products?
- How much material and energy does each require within our company as well as over its entire life cycle?
- What are the largest areas of environmental concerns (i.e. material production, product production)?
- How should we design future products based on economic and environmental costs?
- How much of our products' environmental impact is due to our manufacturing process, versus the load induced by our suppliers?

- What activities are incurring the most economic and environmental costs?

<u>Upper Management</u>
Job: Deciding business strategy, product lines, pricing
- Where does non-value added spending occur?
- Does the product prices accurately reflect true costs to our company? To society?
- Currently, what is the environmental impact of each of our products?
- Overall, is our business reducing its environmental impact? Show yearly trends in each measured indicator.

Right now, our emphasis is on supporting the plant supervisor, manufacturing engineer, and upper management. We feel that at the worker level, traditional control systems are very well established and suited for unit process optimization. Getting the data and production information to higher-level management in a timely and effective format that supports economic and environmental decision-making is what we focus on. However, we are starting to look deeper into how exactly we could help the design engineer once we have production data.

DEVELOPMENTAL QUESTIONS ENCOUNTERED AND LESSONS LEARNED

As stated, our first prototype will go online very soon (before publication of this article). During our development process, we encountered several issues that we had to resolve. Some may look familiar to many readers, but we felt it would be worthwhile to list some of the more notable ones. We believe that anyone who is interested in this kind of work and real implementations will encounter similar issues. Some of the issues that we had to resolve were:
- What are the characteristics of the potential users of the dashboard, and how do they help us define the functions that the dashboard must support?
- What are the implications of the time quality of data when designing the system to support queries?
- How close to "real-time" is necessary and achievable for sensor measurements in an information network like the one underlying the dashboard?
- To what extent do we need to gather real-time sensor data (i.e., sub-meter processes/activities) in order to support the updating of drivers and amounts in an Activity Based Costing model? What sensors do we need?
- What process performance indicators (for efficiency, effectiveness) make the most sense to measure for decision making in conjunction with ABC model data?
- What industrial-strength information systems technologies (e.g., PLCs) are best suited for the accumulation of sensor data? What are the biggest bottlenecks for data-transfer speed? How useful are the latest technology features (web-enabled PLCs, bus networks, etc.) to the needs we have?
- How should databases be configured to support queries? Should multiple databases be used? How do we communicate between them?
- Is Java an appropriate solution for the data-passing requirements of our system? What are its limitations (speed, flexibility, etc.)?
- How do we integrate production, cost, and accounting data? Automatically or manually? How do we set up procedures to do so manually?
- What is the best way to show various types of information onscreen in a browser format?
- What technologies exist to make the dashboard GUI more easily configurable for process changes or introduction into new activities? (Cold Fusion™, Visual Café™, etc.)

We are still struggling with some of the issues, but we have learned some valuable lessons. Some of these are:
- Communication between heterogeneous PLC and data networks requires programming experience in multiple languages, and conversion libraries between them are often scarce.
- Legacy factory networks (in our case, Token ring) require serious hardware investments and may even interfere with data sharing between plants.
- Firewalls can pose a serious obstacle to the sharing of data within a corporation.
- Specialized sensor bus networks like Profibus are not necessary (or even available) for most of our sensing needs. They are better for control functions, which we currently do not support.
- Web-enabled PLCs perform poorly and will soon be phased out and replaced by PC based cards that do the same thing.
- There will be some process variables that cannot be feasibly measured automatically in real-time. A detailed study of the process is critical in the early stages.
- High-level overall production rates and "real-time" data may not make sense if there is a large variety in which products go through which lower level activities.
- Java is largely unsupported by high-end PLC vendors. Microsoft-related programming environments dominate here, as do proprietary, vendor specific software packages for the PLCs.
- Cold Fusion™ is an excellent platform for easy web-based queries that yield tables. This removes a lot of computational effort from the client.
- Scheduling sensor installation around a fully ramped-up production process requires full support by the production engineers if you want your monitoring to be implemented in reasonable time.

Especially the latter point is important. Without good support from the production engineers, implementations are much, much more difficult. Fortunately, we did not have this problem.

CLOSURE

In this paper, we illustrated our approach and progress to building a prototype system for integrated economical and environmental performance monitoring. In our opinion, our approach has three unique elements:

- Rather than starting from an environmental assessment or management approach and trying to tie this with traditional business practices, we approach the problem from the opposite side. We build upon existing and (for companies) familiar cost management principles and to extend those into environmental management, providing familiar starting points for companies seeking to improve their environmental performance.
- In order to alleviate the burden of data and information gathering, we use an increased number of sensors in manufacturing. In addition to having conventional sensors for process control, we add sensors for measuring energy and material consumption and waste/emissions.
- We use the internet and web-based technology for transferring and viewing the data. We foresee web-browser based tools ("dashboards") for monitoring available at the factory floor level, at the plant-manager level, as well as at the corporate level.

Our first prototype will go online in November 2000, but already the modeling exercise alone has yielded some new insights for Interface regarding their production and its economic and environmental impact.

ACKNOWLEDGMENTS

We gratefully acknowledge Interface, Inc. for their contributions and support. We also acknowledge the support of NSF Grants DMI-0085253 and DMI-0086762.

CONTACT

For more information, please contact the corresponding author Dr. Bert Bras, Woodruff School of Mechanical Engineering, Georgia Institute of Technology, Atlanta, Georgia 30324-0405, USA.

REFERENCES

1. Emblemsvåg, J. and B. Bras, *Activity-Based Cost and Environmental Management – A Different Approach to ISO 14000 Compliance.* Kluwer Academic Press, New York, 2001.
2. Emblemsvåg, J. and B. Bras, *Industry experiences with Activity-Based LCA.* Society of Automotive Engineers, Paper No. 2000-01-1464, 2000.

Development of the Chemical Recycling Technology of Glass Fiber Reinforced PA6 Parts

Kaoru Inoue and Yuichi Miyake
Toyota Motor Corp.

ABSTRACT

Recently , the plastic material is positively introducing for automotive parts due to the Needs of vehicle weight reduction and cost saving. On the other hand, the countermeasure for scrapped car is a big subject to need to consider as a car maker. Therefore, the development of recycling technology for plastic parts has been necessary. In this study, we tried to develop recycling technology for glass fiber reinforced Polyamide6(PA6) which is applied to various automotive parts like an air intake manifold. As a recycling technique, we focused on the chemical recycling which can reclaim raw material of PA6(ε- caprolactams) from the post-consumer automotive parts. The chemical recycling we selected can be put on a higher priority because it has possibility to utilize the limited resource repeatedly. As a result, we could retain high purity of ε- caprolactams using our following two techniques which make possible to recycle Polyamide 6 materials. One is to separate PA6 from glass fiber. To optimize pressure is important after decreasing the resin viscosity with optimized organic acid. Another is to depolymerize PA6 continuously using phosphoric acid which separated the glass fiber. Recycled PA6 which we synthesized from using ε-caprolactams from the above method showed same level of physical property as original PA6.

INTRODUCTION

More and more interior and exterior automobile parts are being made of plastic in order to reduce weight and lower cost. In recent years, the use of plastic for functional parts such as intake manifolds has increased dramatically. The aim being to reduce weight even further and to cut down on the number of parts through modularization (Figure 1). For a functional part such as the intake manifold, a material must be chosen that can withstand the severe usage environment. Due to its superior heat resistance and balance of physical properties, glass fiber reinforced PA6 material has been the main choice for such parts.(Figure 2).

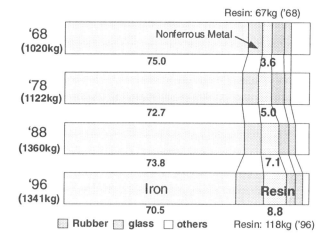

Figure 1. The change of material construction ratio

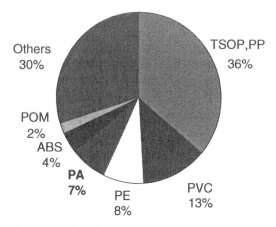

Figure 2. The classification of using resin material

While the increasing use of plastic parts has brought with it a number of advantages in terms of reduced weight and lower cost, it also presents the unavoidable problem of what to do with these parts when a vehicle reaches the end of its useful life (Figure 3). For this reason, it has become necessary to develop a recycling technology for PA6 plastic parts.

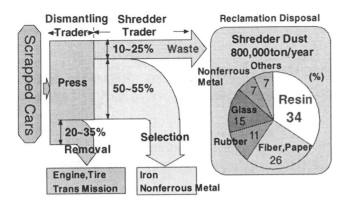

Figure 3. The resources and reclamation disposal of Scrapped cars

In the research reported in this paper, a chemical recycling technology that breaks down the plastic into monomers for reuse as a chemical raw material was studied as one method for recycling the glass fiber reinforced PA6 material that is used for the plastic intake manifold.

SELECTION OF CHEMICAL RECYCLING METHOD

SCREENING OF RECYCLING METHODS

Table 1 describes four recycling methods and their characteristics. The material recycling, thermal recycling, and reuse methods all present problems, from limitations on the parts that can be processed to the impossibility of using the resources repeatedly. Therefore, for environmental reasons, the chemical recycling method of chemically breaking down the material and reusing it as a chemical raw material was chosen because it makes it possible to recycle the material semi-permanently.

Table 1. The change of material construction ratio

Method	Description	Characteristics
Chemical recycling (Feed stock)	Decompose chemically, use as chemical raw materials	•semi-permanent •Highly difficult in terms of both technology and cost
Material recycling (Mechanical)	Melt same type or different. Types of materials and reuse.	•Applicable part types are limited. •Limits on repeated use
Thermal recycling (Energy recovery)	Reused as thermal Energy	•sorting not normally required •resources can not be used repeatedly
Reuse	Reuse as parts	•Applicable part types are extremely limited.

SUITABILITY OF PA6 FOR CHEMICAL RECYCLING

Table 2 shows the suitability of the main engineering plastic materials for chemical recycling. It can be seen that the PA6 material is composed of a single type of monomer (ε-caprolactam). Also, because its polymerization is a balanced reaction, chemical decomposition is easy, making the material especially suitable for recycling.

Table 2. Suitability for chemical recycling

Plastic	Main constituent Monomer	Polymerization Reaction	Polymerization Monomer recovery
PA6	ε-caprolactam 1 type	Reversible (Ring-opening Polymerization)	Easy
PA66	Adipid acid Hexamethylenediamine 2 types	Reversible (condensation Polymerization)	Difficult
PBT	Terephthalic acid Butanediol 2 types	Reversible (condensation Polymerization)	Difficult

CLARIFICATION OF ISSUES

The basic process flow of the chemical recycling of the plastic intake manifold is shown in Figure 4. After the steel inserts and other non-recyclable parts are removed, the plastic intake manifold is washed and pulverized. Next, the pulverized material is diluted and melted to separate out the plastic component, which is then depolymerized to obtain raw lactam. The raw lactam is further refined to extract the ε-caprolactam, which is then used in ring-opening polymerization to synthesize PA6, the material used for the intake manifold.

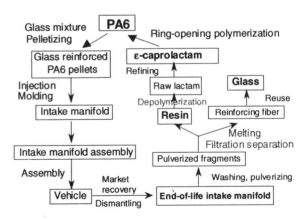

Figure 4. Chemical recycling process flow for PA6 Intake manifold

In order to efficiently recover the lactam that is the raw material for the PA6, two technologies are critical. The first is technology to separate the plastic component from the glass fiber reinforced PA6 material, and the second is technology to increase the speed of the depolymerization reaction that breaks down the plastic component.

STUDY OF TECHNOLOGY TO SEPARATE PLASTIC COMPONENT

SOLUTION VISCOSITY REDUCTION FOR SEPARATING PLASTIC COMPONENT

Technology that uses filters to separate the plastic component from the reinforcing glass fibers was studied. However, it became clear that the solution viscosity of the PA6 material with 30% glass fiber reinforcement was so high, approximately 500 Pa•s, that the solution could not be easily filtered. The laboratory device shown in Figure 5 was designed to address this problem and a study of solution viscosity reduction for separating the plastic component was conducted (Figure 6). The results of this study made clear that adding six parts of acetic acid as a diluting agent to one part of PA6 and heating the mixture to 250°C could lower the solution viscosity to 20 Pa•s. However, the amount of plastic component that can be extracted by reducing the solution viscosity alone is limited to 70%.

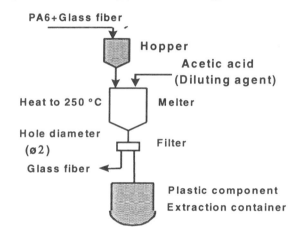

Figure 5. Laboratory device for melting and viscosity Reduction

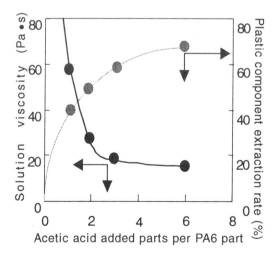

Figure 6. Result of solution viscosity reduction study

COMPRESSION SEPARATION OF PLASTIC COMPORNENT

Because there is a limit on the amount of the plastic component that can be extracted by reducing the viscosity of the solution, the addition of a compression process was studied and the laboratory device shown in Figure 7 was designed. The effect of the compression pressure when this device is used is shown in Figure 8. As the graph shows, when the solution is heated to 250°C and compressed under a pressure of 0.6 Mpa, the glass fiber forms a residue, and approximately 80% of the plastic component can be separated and recovered.

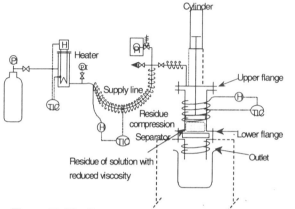

Figure 7. Plastic component compression separation

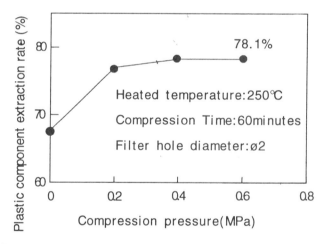

Figure 8. Effect of compression pressure

STUDY OF TECHNOLOGY TO INCREASE DEPOLYMERIZATION REACTION SPEED

STUDY OF CATALYST IN DEPOLYMERIZATION REACTION

The following is the depolymerization formula for PA6. The depolymerization reaction is a process in which a catalyst is used to obtain from the PA6 material the lactam that is its raw material.

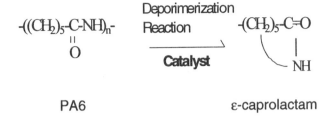

PA6   ε-caprolactam

Three depolymerization methods are available: Wet method, Dry method and High temperature/High pressure hydrolysis method. The wet method was chosen for its simplicity and the small scale of the equipment required. It is a method that uses steam as a carrier gas to extract the lactam that is the raw material for PA6 from the solution. Depolymerization catalysts were studied on the assumption that they would be used in this depolymerization method (Table 3) .

Table 3. Study of catalyst in PA6 depolymerization reaction

Substance name	Device corrosion	Reaction Speed	Cost	Overall evaluation
Phosphoric asid	O	O	O	O
Sulfuric asid	X	O	O	X
Organic asid	△~O	△~O	△	△
Sodium Hydroxide	△	△	O	△

This table shows the results of a comparison of phosphoric acid, sulfuric acid, organic acids, and sodium hydroxide as depolymerization catalysts. The results make clear that phosphoric acid is the best choice, because of its limited corrosion of the device and its fast reaction speed.

STUDY OF DEPOLYMERIZATION REACTION TIME

Figure 9 shows the relationship between the depolymerization reaction time and the percentage of lactam extracted. For the purposes of this study, three materials were compared: unreinforced PA6, PA6 with 30% glass fiber reinforcement, and intake manifold material recovered from the market. For all of the materials, the percentage of lactam extracted increased as the depolymerization reaction time became longer, with approximately 90% of the lactam being extracted when the reaction lasted for six hours.

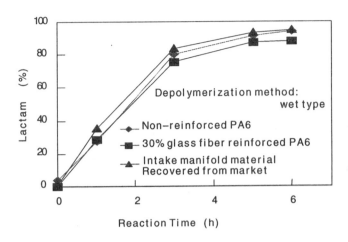

Figure 9. Effect of depolymerization reaction time

LABORATORY STUDY vs PILOT PLANT STUDY

CORRELATION WITH LABORATORY DEVICE

Based on the results of the laboratory studies, a study was conducted of continuous separation and depolymerization in a pilot plant, using pulverized intake manifolds recovered from the market. The results showed that the plastic component separation process was able to separate approximately 80% of the plastic, and the depolymerization process was able to recover approximately 90% of the lactam. This indicates that the same level of processing that is possible in the laboratory can be achieved in the pilot plant.

QUALITY OF RECYCLED LACTAM

In order to confirm the quality of the lactam that was obtained at the pilot plant, analyses were carried out using both gas chromatography and liquid chromatography (Table 4). As can be seen from the table, impurities are present in the recycled lactam in small amounts.

Table 4. Lactam quality

	GC purity(%)	HPLC impurities (ppm)
New lactam	>99.99	Not detected
Recycled Lactam	>99.99	47 ppm

STUDY OF PHYSICAL PROPERTIES OF RECYCLED MATERIAL

In order to determine whether or not the impurities in the lactam that is obtained using the technology described above influence the physical properties of the material, a material evaluation was carried out using PA6 material that had gone through the refinement and polymerization processes. The results are shown in Table 5. They show that the recycled PA6 material that was synthesized from the recycled lactam has the same level of physical properties as the new material.

Table 5. Mechanical Properties of 30% glass fiber compound product

Mechanical Properties		PA6 new Material	PA6 recycled Material
Tensile Properties	Strength (Mpa)	187	188
	Elongation (%)	3.5	3.6
Izod impact (J/m)		151	151
Bending Properties	Strength (Mpa)	244	247
	Modulus (Mpa)	8558	8511

CONCLUSION

(1) A chemical recycling technology was developed for glass fiber reinforced PA6 plastic parts such as the intake manifold.

(2) The salient technical points are that the technology separates the plastic component of the material and the glass fiber by means of melting and filtering, and that depolymerization technology is applied to the separated plastic.

(3) A heating and compression separation method was developed for the filtering technology, and a wet method using phosphoric acid was developed for the depolymerization technology.

(4) A recycled PA6 material with the same level of material properties as new material was created using ε-caprolactam that had been obtained by continuous separation and depolymerization in a pilot plant.

CONTACT

Kaoru Inoue: Material Engineering Div.2, Toyota Motor Co., E-mail: kaoru@inoue.tec.toyota.co.jp

REFERENCES

1. Yoshinori, H.: "Plastics," vol.47,No.7,P41 (1996)
2. Takashi, S.: "Plastic Age," 1997,Special Issue (1997)
3. Osamu, F.:"Polyamide Resin Handbook," Nikkan Kogyo Shimbun (1988)
4. Haruyasu, S.: "Roka Wa Kataru," Chijin Shokan (1994)

2001-01-0697

Size Reduction and Material Separation Technologies for the Recovery of Automotive Interior Trim ABS and PP Plastics from Dismantled End-of-Life Vehicles: Preliminary Research into Continuous Processing

Michael M. Fisher
American Plastics Council

Michael B. Biddle and Chris Ryan
MBA Polymers, Inc.; Metech International

ABSTRACT

The life-cycle benefits of plastics in automotive applications are significant and continue to expand. Achieving cost-effective recycling of interior plastic components obtained from end-of-life vehicles (ELVs) remains a formidable technical and economic challenge worldwide. The American Plastics Council (APC) has pioneered large-scale process development work toward advancing environmentally and economically sound recovery technologies for plastics from ELVs. Groundbreaking studies have focused on the recovery of ABS and PP interior trim plastics. A combination of dry and wet processing of over 50,000 pounds of interior trim parts obtained through selective dismantling has been completed and streams of ABS and PP isolated. This work included the evaluation of fast mid-infrared plastics ID technology to facilitate the separation and purification process. Both sorted and unsorted trim was processed. The overall program objective was to obtain both of these recovered plastics in acceptable form for possible reuse in automotive or other applications. Detailed information on process equipment and plant design along with process strategies and detailed material balance data is presented. This work quantified the limitations of density separation to achieve clean streams of unfilled PP, filled PP, and unfilled ABS interior trim plastics from end-of-life vehicles.

INTRODUCTION

This paper presents the results of preliminary research funded by the American Plastics Council (APC) to investigate the recovery and recycling of acrylonitrile-butadiene-styrene (ABS) and polypropylene (PP) automotive interior trim plastics from end-of-life vehicles (ELVs). The overall program objective was to obtain both of these recovered plastics in acceptable form for possible reuse in automotive or other applications. A combination of dry and wet processing of trim parts obtained through selective dismantling of ELVs was explored for both sorted and unsorted mixed interior trim. In some of the trials, rapid mid-infrared plastics identification technology was used to assist in the sorting. A total of over 50,000 pounds of post-use automotive interior trim were processed in this study.

This research was carried out by wTe Corporation (Bedford, MA) under contract to APC using the Multi-Products Recycling Facility (MPRF) in Dorchester Massachusetts. The principal investigator was Chris Ryan of wTe Corporation. Dr. Michael B. Biddle of MBA Polymers, Inc. (MBA) and Dr. Michael M. Fisher of APC provided technical assistance.

This paper covers both Phase I processing of ABS and PP interior trim using the original MPRF configuration and subsequent Phase II studies after reengineering several MPRF processing steps.

PHASE I TEST PROGRAM

OVERALL PROCESS DESCRIPTION - Figure 1 presents a block flow diagram of the size reduction and material separation line as originally installed at the MPRF for the Phase I APC trials. This system was designed to shred whole plastic parts using a modified shear shredder into a nominal 5/8" chip size. The shredded material was screened to remove fines and oversized chips, passed over a drum magnet to remove ferrous materials, and then either boxed for subsequent processing or fed directly to the downstream unit operations to recover non-ferrous metals and plastics.

These operations included an air classification system to remove paper, fiber, and other light materials from the shredded product. Following air classification, material entered the density separation and cleaning process. This portion of the system consisted of two spiral classifiers (sink-float tanks) arranged in series, followed by a rinse/wash system configured to recover a clean single-resin plastic product. The spiral classifiers were used to separate the plastic materials by density, using salt-water baths whose concentration could be adjusted to target products of different densities.

The system was normally configured as shown in Figure 1 to produce light, middle, and heavy density cuts; with the middle cut typically being targeted as the product stream. In this operating mode, the floating materials from the first classifier ("CI Lights") and the sinking materials from the second classifier ("CII Heavies") were removed from the target material stream, and the middle cut (product) was sent directly to the wash system which used hot water and a detergent for cleaning. The wash step was followed by a rinse step, dewatering screen, and spin dryer. The light and heavy cuts from the classifiers were either discarded as process residue or saved for subsequent processing.

The process line also contained a mineral jig, located between the two classifiers, which was used as an additional metal removal step to eliminate mixed non-ferrous metals, including wiring harnesses, molded-in metal fasteners, or non-ferrous metal parts from the plastic product stream.

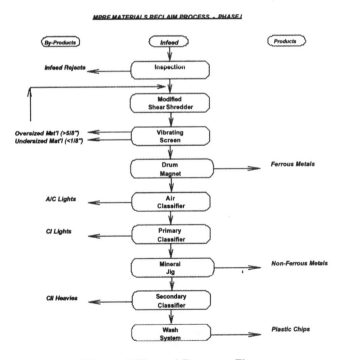

MPRF MATERIALS RECLAIM PROCESS - PHASE I

Figure 1 Phase I Process Flow

DETAILED PROCESS DESCRIPTION - A description of each unit operation used in the wTe-APC process is presented in the following section. The function of each

piece of equipment is described along with additional information about actual operating experience.

Size Reduction System - The size reduction step is employed to reduce the incoming materials to a uniform particle size and to liberate various attached materials (metals, plastics, paper, foam, etc.) from the plastic parts so that the downstream separation equipment can separate the various fractions from each other. A 100 HP, hydraulically driven 52" wide shear shredder was used in the original MPRF process due to its ability to handle thick pieces of tramp metal. The shredder was modified to provide a 1/2" cutter spacing across one half of its shaft length vs. the 1-1/2" spacing that was installed on the original machine. The cutters along the remaining length of the shaft were left at their 1-1/2" spacing. This was done in order to retain some coarse-cut capability in case it was needed for larger, bulkier objects. A vertical divider plate was installed to separate the coarse-cut from the fine-cut side of the machine. In addition, a custom size control screen with 5/8" openings was fitted to the underside (discharge) of the shredder to provide a more uniform particle size and increase retention time to improve the liberation of materials.

The shear shredder was also fitted with a mechanical device to assist in feeding material through the shredder. This feeder was a steel platen mounted so that it pivoted on a shaft which in turn was connected to an oscillating mechanical drive. When actuated by an operator feeding material onto the shredder infeed conveyor from ground level, the drive turned the shaft, causing the platen would swing down over the top of the shredder shafts and force bulky materials into the cutter teeth and through the shredder. The device was used as a feed assist for large, tough items that tended to ride on top of the shredder shafts without being grabbed by the teeth such as large flat pieces of automotive interior trim. Although this feeder was custom fabricated as a retrofit for this shredder and application, similar feeders are generally available as options from most size reduction equipment manufacturers.

Secondary Size Control Screen - With the modified shredder configuration, some material bypassing from the fine-cut side to the coarse-cut side was experienced, producing some oversized material. A double-decked vibrating screen was installed directly below the modified shear shredder as a secondary size control device to catch oversized material (+ 5/8"). This material was typically collected and fed back to the process infeed for re-shredding. As each box of material was re-fed, its weight was logged in order to monitor the relative effectiveness of the shredding step for different materials. The recycle rate was recorded as the total amount of material re-fed to the shredder divided by the total material initially fed to the process. During the Phase I trials, a wide range of values were recorded, exhibiting an average value of 58%, and ranging from 11% to 84%. For some materials, the shear shredder was operated with the coarse cutters exposed, which contributed to the higher recycle rates but which allowed problematic tougher parts to be pulled through the machine. In all cases, the lower deck of the screen was used to remove fines (-1/8") from the process before the material was fed to the downstream separation equipment.

Ferrous Metals Removal System - A standard drum magnet was used as a means of separating the ferrous materials from the processed materials. Material was fed onto the surface of a rotating non-magnetic drum. Magnetic materials cling to the drum as its surface passes over an internal fixed magnet. After passing the magnet, the ferrous materials drop from the drum surface into a discharge chute that deposits the material into a collection bin.

Air Classification System - A system employing counter-current airflow was used to remove liberated paper, foam, film, and additional fines from the shredded material prior to feeding it to the wet density separation process. This step was employed to reduce interference with the rigid plastics in the sink/float process by extraneous fluff, which can be more difficult to wet and tends to build up on the surface of the bath.

Density Separation System - The density separation system consisted of two spiral classifiers, arranged in series, each containing a liquid bath used to separate materials by flotation due to density differences. By using different solutions of water and salt, the specific gravity (ratio of measured fluid density to water) of the separation media used in the two classifiers was set to allow different materials to float and sink, thereby facilitating recovery of the targeted resin. Material from the size reduction system was fed by gravity from a conveyor directly into the first classifier's bath *primary classifier,* where heavy material would sink and be conveyed by a submerged screw conveyor to the heavies discharge. Light materials, which floated to the surface, were conveyed by surface currents to the rear of the vessel where they were discharged in a liquid slurry into a discharge chute and then onto a de-watering screen. The chipped materials were separated from the liquid by the de-watering screen and sent either to a mechanical dryer prior to being discharged from the process or to the next unit operation in the process *secondary classifier.* The liquid collected by the de-watering screens was collected in dedicated tanks and pumped back into the appropriate classifier vessel by a circulating pump.

SOURCING OF THE PLASTICS FOR PHASE I - The automotive trim processed during Phase I of the trials was collected by wTe on behalf of APC. Rigid, mixed color plastic automotive interior trim pieces, such as door panels, pillar moldings, and kick panels were recovered from auto dismantling yards in the Greater Boston area for processing at the MPRF. The selected parts were molded primarily of ABS and PP. During collection of the material

used for the Phase I trials, minimal pre-screening or cleaning of the collected parts was employed in the field. However, the Phase I material did undergo extensive manual pre-cleaning and sortation prior to installation of the MPRF process line as a part of that collection and characterization effort.

During the Phase I testing, approximately 20,000 pounds of material was processed to recover clean PP and ABS product streams. During this series of trials, three separate groups of automotive interior trim were processed. The largest group (approximately 11,000 pounds) consisted of mixed ABS and PP parts, with the two smaller groups consisting of pre-segregated quantities of PP (7,000 pounds) and ABS (2,000 pounds). Pre-sorting was done by visually inspecting the parts for molded in resin identification labels. All of the materials had been manually pre-cleaned of most contamination (carpeting, insulation, clips, fasteners, electronic components, etc.) prior to processing.

MATERIAL PROCESSING – SPECIAL CONSIDERATIONS

This section documents some of the detailed observations made while running selected Phase I materials. They provide insight into some of the operating issues experienced that were used to define the capabilities and limits of the process as originally installed. This information was used to establish some of the basic design criteria used for the system upgrades implemented in Phase I of the testing program. The observations and materials presented in this section were chosen to cover a complete range of operating issues addressed by the test program including: impact of part design and plastic type on size reduction system operation, impact of fillers in plastics on density separation system operation, and impact of structural foam on density separation system operation

Test M131-2 - PP and ABS from Mixed Automotive Interior Trim - In this test run, over 11,000 lbs of automotive interior trim was processed in a single-pass run to recover PP and ABS products. Water was used as the bath media in the primary classifier to recover PP, and a salt solution at a density of 1.08 g/cc was used in the second classifier to recover ABS. The size reduction capacity of the system was lower than originally anticipated. This was attributed primarily to the large part size (> 24" dimension) and light, bulky nature of some of the pieces, which tended to hang up on the 24" wide skirted infeed conveyor, and within in the shredder chamber itself, where many of the large, flat pieces, particularly whole door panels, tended to lay flat over the shredder cutters, preventing them from being pulled into the nip of the cutters. Attempts were made to increase throughput by pre-shredding in a hammermill, but it also proved unable to handle the large pieces without bridging in the small infeed hopper. Finally, the steel baffle plate which was normally installed to cover and de-activate the course-cut side of the shredder was removed to allow the

1-1/2" cutters to be exposed. While this resulted in an increased production of oversized material (> 5/8") requiring additional re-processing, the larger cutters (which have a higher tooth height) improved the ability of the machine to grab material and pull it through the shredder, improving the overall throughput. Visually, it appeared that more than half of the material was still being pulled through the fine cut side of the machine, even with the coarse cutters fully exposed. Exposure of the coarse cutters resulted in an improvement in throughput from an average of 274 lbs/hr during the first four days of processing attempts to 474 lbs/hr over the final five processing days. It was determined that a two stage size reduction system should be evaluated as a way to achieve higher throughput when mechanically size reducing this material.

Because of the throughput problems encountered during size reduction of this material, operators concentrated exclusively on the size reduction operation and staged the shredded material for subsequent processing through the wet separation system. Air classification was performed as a separate step prior to the wet separation process. This was done both to accommodate the throughput differences noted between the dry and wet processing systems and also to provide a more even infeed rate to the air classifier which is sensitive to this operating parameter.

The wet process system was configured to process the mixed interior trim in a single pass. The capacity of the density separation system appeared to be limited by the available surface area of both classifiers. The primary classifier, normally used only to remove a lesser amount of non-target materials, was overloaded due to the high volume of PP in the stream. The feed rate was regulated to prevent excessive buildup of material on the water surface where the floating material was transported to the lights discharge. Plain water was used to increase the sink rate of the heavy plastics. The fluid flow in both classifiers was kept at a minimum to prevent excessive carryover of heavy chips (sinkers) into the lights discharge, in an effort to maximize separation performance.

Another operating problem experienced during processing of this material was jamming of the rotary airlock at the discharge of the vac-loader, which pneumatically conveys material into the main air classifier. This was experienced on several occasions. Upon inspection, it appeared that a buildup of fluff and fibrous material from the insulation and padding used in these parts was occurring. The buildup between the open ends of the rotor vanes and the unit's housing caused increased friction, resulting in jamming of the rotor. Several times during the run, processing through the separation line had to be stopped to allow the airlock to be cleared. The air classifier manufacturer was contacted to assist in providing a solution to this problem, which ultimately involved replacement of the open-ended rotor with a closed drum type of unit. This also illustrated

a need for improvements to the size reduction device to provide better liberation of the fluff materials from these plastic parts.

Test M131-3 - PP from Pre-sorted Automotive Interior Trim - Similar low throughputs were experienced with pre-sorted polypropylene trim. As with previous processing of mixed PP and ABS interior trim, the steel baffle plate normally installed to cover and de-activate the coarse cutter side of the shredder was left off in order to expose these cutters. While this resulted in an increased production of oversized material (> 5/8") requiring additional re-processing, the larger cutters (which have a higher tooth height) increased the machine's ability to grab material and pull it through the cutters, improving the overall throughput. In spite of this measure, throughput, at approximately 200 lbs/hr, was unacceptably low. The pure polypropylene stream appeared to be even more difficult for the shredder to grab and break up. This was attributed in part to it being less brittle than the ABS containing materials previously processed, therefore requiring more positive cutting action.

A mechanical ram feeder (pivoted steel plate), had been installed on the shredder prior to this run and was tested on this material. This device was used to help push bulky materials that tended to bridge on top of the shredder cutters through the machine to improve throughput. Using this method, an average throughput rate for the entire run of 202 lbs/hr was obtained vs. 238 lb/hr for a mixed ABS/PP trim stream. However, the recycle rate was less for the mixed stream, (34% for the PP vs. 51%) which was attributed to the use of the ram feeder assist device, which helped to push more of the material directly down through the fine cut side of the shredder on the first pass. Use of the ram feeder proved essential for processing the less brittle PP trim through the size reduction system. More importantly, it freed a second operator from having to manually assist the material through the size reduction system, allowing the size reduction line and the separation process line to be run simultaneously.

The capacity of the density separation system was limited by the available surface area of the secondary classifier. Due to manual presorting, the percentage of target resin (PP) in the stream was unusually high. To accommodate this, the feed rate was reduced to prevent excessive buildup of material on the water surface of this classifier, where the floating material is transported to the lights discharge. Plain water was used to maximize the sink rate of the heavy plastics. The fluid flow was kept at a minimum to prevent excessive carryover of heavy chips (sinkers), in an effort to maximize separation performance.

Another unique operating problem was noted during processing of this material. This material tended not to flow into the infeed chute of the de-watering spin dryer (between the mineral jig and the secondary classifier). The rough, soft surface of the chips seemed to cause flow problems. The infeed chute of this spin dryer contains a perforated plate over which the material slides, and the PP chips tended to hang up on the plate, causing pluggage. The perforated plate was replaced with a smooth slide, which allowed material to be fed without plugging. Overall, the wet separation system was operated at a reduced average throughput on this material in order to accommodate this problem, the limited chip carrying capacity of the secondary classifier, and the low production rate of the primary size reduction system.

Test M131-4 - ABS from Pre-sorted Automotive Interior Trim - The size reduction system configuration for this material was identical to that used for the pre-sorted PP stream (Test M131-3), and throughput rates experienced were similar, which indicated that the influence of material type on low throughput was smaller than originally thought.

The throughput recorded for the density separation system was significantly higher than what had been found for other materials in earlier studies, even approaching 1.6 times the rate obtained during processing of a relatively pure ABS stream from telephone housings. This was attributed to the fact that a surfactant was used in both classifiers which improved particle wetting, and therefore, separation performance at higher throughputs, particularly in the primary classifier. The ABS trim was also observed to flow more readily through transition points where flow problems had been previously observed while processing polypropylene and mixed trim, contributing to reduced throughput. The fluid flow was kept at a minimum to prevent excessive carryover of heavy chips (sinkers) into the ABS product and to maximize separation performance. Like the other interior trim runs, a buildup of fibrous material at several points in the water handling systems (dewatering screens and sump pumps) was noted indicating the additional liberation of fibrous material during wet processing, confirming the need for its removal at another location in the system.

Table 1 shows the detailed baseline processing information and mass balance data for the three trials described above. Targeted plastic yields ranged from 73 to 77 percent.

PHASE II OBJECTIVES - The objectives of Phase II were to implement and test a series of process modifications designed to increase processing capacity and to overcome some of the operating instability observed while conducting Phase I testing. The changes described in this report were implemented in two stages; the wet separation line upgrades were completed first, and the modifications to the size reduction system were completed six months later. Following the completion of those upgrades, a series of preliminary processing trials was conducted which were designed to evaluate the performance of the individual unit operations that had been upgraded.

After the process modifications and preliminary test work was completed, a series of fourteen full-line processing trials were performed to generate material balances, material samples, and monitor system operation as had been done in Phase I. These results were compared with the earlier data and used to further define the capabilities (and limits) of the process.

DESCRIPTION OF MPRF PROCESS MODIFICATIONS Phase II process changes included increasing the size of the primary classifier bath/settling chamber and replacement of the secondary classifier with a larger unit to provide more stable material flows and longer material residence times at higher throughputs in the density separation system. The classifier bath circulation systems were improved through the addition of spray nozzles and filtration systems to provide more controlled surface flows for improved material handling.

The other major modification made to the process was the installation of a rotary grinder as a size reduction device. This machine had been identified as the most likely to meet the needs of this application as the result of a world wide technology search effort that was part of an APC-sponsored program undertaken to identify alternative size reduction methods for durables. As a part of this overall system upgrade, the original shear shredder was restored to its original coarse-cut configuration (1-1/2" cutters) and kept in the line for use as a primary size reduction device for selected materials containing excessive metal contamination such as steel frames and heavy duty mechanical parts like small motors, drives, etc. that could not be handled by the rotary grinder. A 36" wide cross-belt magnet was incorporated into the shear shredder material discharge conveyor to remove the large pieces of liberated ferrous metals from these parts so that the remaining materials could be fed continuously into the rotary grinder. In addition, several intermediate material handling improvements were incorporated into the process line.

Table 1 – Phase I Test Results for Automotive Interior Trim

Test Report No: Target Plastic: Description:	M 131-2 ABS/PP Int Trim	M 131-3 PP Int Trim	M 131-4 ABS Int Trim
Size Reduction System Material Balance			
Infeed Material	11,138	7,160	1,968
Rejects - Manual Inspection			
Oversized Screened Material at End of Run (>5/8")	0	0	0
Ferrous Metal	90	64	20
Air Classifier Lights	528	398	32
Q/C Sampling (Size Reduction)	0	0	0
System Losses (Size Control Screen Fines & Spillage)	800	338	112
Wet Process System Material Balance			
Infeed to Wet Process	9,720	6,360	1,804
Primary Classifier Lights	5,076	n/a	102
Mineral Jig Heavy Fraction	368	4	16
Secondary Classifier Heavy Fraction	746	134	82
Q/C Sampling *	240	92	24
System Losses	80	906	78
Target Plastic Product Yield (includes wet line Q/C)	8,526	5,224	1,502
Production Summary			
Size Reduction System Yield (lbs into wet process/lbs into size reduction)	87%	89%	92%
Wet Process System Yield (target resin out/lbs into wet process)	88%	82%	83%
Overall Yield - Target Plastic (lbs target resin/lbs into size reduction)	77%	73%	76%
Size Reduction Process			
No. of Boxes In	107	93	23
No. of Product Boxes Out	11	11	2
% Volume Reduction (Product)	90%	88%	91%
Equipment Configuration	Std	Std	Std
Process Material Throughput	238	202	198
Recycle Rate (Lbs Re-run/Lbs In)	51%	34%	57%
Processing Hours (Mat'l Feed Time)	46.8	35.4	9.9
Total Hours (Including Setup & Downtime)	72.0	48.0	16.0
Utilization Factor (Processing/Total Hrs)	65%	74%	62%
Wet Separation Process			
Process Material Throughput (Lbs/Hr)	508	277	908
Processing Hours (Mat'l Feed Time)	20.2	21.1	2.2
Total Hours (Including Setup & Downtime)	40.0	44.0	8.0
Utilization Factor (Processing/Total Hrs)	50%	48%	27%
Classifier Bath Densities			
First Pass:			
C1 Bath Density (g/cc)	1.00	n/a	1.00
C2 Bath Density (g/cc)	1.08	1.00	1.08
Second Pass:			
C1 Bath Density (g/cc)			
C2 Bath Density (g/cc)			
Summary			
Yield (% of whole parts fed into process)			
Target Plastic	77%	73%	76%
Ferrous metals	1%	1%	1%
Non Ferrous Metals	3%	0%	1%
Other	19%	26%	22%

Another major Phase II improvement for the MPRF was the addition of a Bruker P/ID 28 mid infra-red resin identification system. The unit identified over 30 plastic types within 5 seconds or less, and was used to pre-segregate interior trim samples prior to mechanical processing to assess how this would benefit product quality. The instrument was also used on selected processing trials to monitor density system performance by analyzing grab samples of plastic chips during production to monitor resin types present in the product.

OVERALL PROCESS DESCRIPTION - Figure 2 presents a block flow diagram of the upgraded system installed at the MPRF for the Phase II APC trials, and an equipment list is provided in Table 2. This process system was used to reduce the incoming material to a nominal 5/8" minus chip size before passing through the various material separation operations. As described earlier, the new system incorporated two separate size reduction devices; the first was the original shear shredder used in Phase I but restored to its original coarse cut configuration. This unit operation was used for materials containing heavy gauge metal. The second unit operation installed was the new rotary grinder. This machine was purchased directly from a German manufacturer, Weima, after testing the machine at their facilities in Europe. A 45 kW (60 HP) machine, a Model WLK 12/45 "Jumbo", was specially modified (additional cutters, heavier duty drive system) by the equipment vendor from its original design as a scrap wood grinding machine to be used for the size reduction of plastics, thus adding the capability to handle bulky parts containing significant amounts of metal contaminants. The machine included an integrated hydraulic ram system which provided a positive feed of bulky parts through a slow-speed cutting system. The process improvements obtained by using this machine included a higher, more stable material throughput requiring less operating labor and a more uniform particle size. As Figure 2 shows, the process line was arranged to allow the rotary grinder to be used as either the primary or secondary size reduction device. For the majority of the interior trim Phase II test runs, it was used as the primary size reduction step.

Following the size reduction step, shredded material was conveyed to the secondary size control screen for removal of oversized (+5/8") and fine materials (<1/8" particle size). Although the double-decked configuration was still used in the Phase II processing, the amount of oversized material produced during size reduction using the rotary grinder was greatly reduced, completely eliminating the need to recycle this material. The +1/8" / -5/8" product discharged from the screen was then passed over the drum magnet, to remove ferrous metals, which were collected in a bin. From the magnetic separator, the non-ferrous fraction was pneumatically conveyed to a surge bin, from which it was metered into the air classification system to remove light materials such as paper, insulation, and foam.

The general arrangement of the upgraded density separation and washing system was essentially unchanged from the process as originally installed. Design improvements were made to individual equipment items which included:

- Increasing the surface area of the primary classifier
- Increasing the screw size and surface area of the secondary classifier (Phase I unit replaced)
- Addition of secondary troughs to both classifiers as a backup heavies catch
- An improved method of material introduction with wetting into the second classifier
- An improved method of bath circulation and filtration to control surface material flows

As the test results described in the following sections demonstrate, the increased surface area and residence time provided by the new classifiers resulted in significant improvements to the density separation system's performance as measured in the form of higher throughputs, higher product recoveries (in selected cases), and lower product contamination levels.

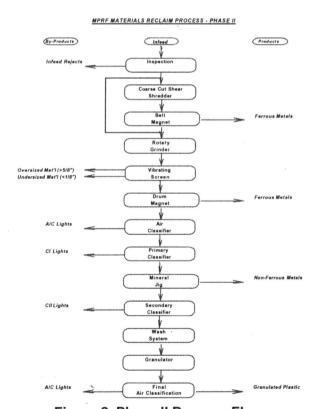

Figure 2 Phase II Process Flow

Table 2 - wTe MPRF Equipment List – Phase II

(Reference to a particular manufacturer in this paper does not represent an endorsement by the American Plastics Council, wTe Corporation, or MBA Polymers, Inc.)

Description	Manufacturer/Source
Primary Size Reduction & Ferrous Metal Removal System	
Infeed Conveyor – 60" Trough Belt	wTe
100 HP Shear Shredder w/ mechanical feed assist,	Saturn
Shredder Discharge Conveyor – 24" Trough Belt	wTe
Belt Magnet	ERIEZ
Primary/Secondary Size Reduction System	
24" Cleated Infeed Conveyor	wTe
60 HP Rotary Grinder	Weima
Rotary Grinder Discharge Conveyor	wTe
Secondary Size Control/Ferrous Metal Removal System	
24" Cleated Infeed Conveyor	
Two-Deck Vibrating Screen	Rampe
Drum Magnet – 14" Dia	ERIEZ
Pneumatic Bin Infeed System (Kongskilde Poly Vac 40)	Kongskilde
Surge Bin w/ Vibratory Feeder	wTe
Air Classification System	
Pneumatic Infeed System (Kongskilde Poly Vac 40)	Kongskilde
Air Classifier System (Kongskilde KF-20)	Kongskilde
Density Separation System	
Slide Belt Infeed Conveyor	wTe
Modified 12" Primary Classifier System (CI)	wTe
36" Dewatering Screen	SWECO
Spin Dryer (CI Lights)	Carter-Day
Mineral Jig	CARPCO
Product Transfer Mechanical Dryer (Intermediate)	Gala
12" Secondary Classifier System (CII)	wTe
Spin Dryer (CII)	Carter-Day
48" Dewatering Screen (2-deck)	SWECO
Wash System	
12" Infeed Screw Conveyor	wTe
Heater Washer System	Martin/Custom
Process Water Heating/Circulating System	wTe
48" Dewatering Screen (2-deck)	SWECO
Spin Dryer (Product)	Gala
Water Treatment System	
Main holding tank	wTe
48" Dewatering Screen (2-deck)	SWECO
Circulating pump system	wTe
Auxiliary Size Reduction Systems (stand-alone equipment)	
(System 1) 60 HP Hammermill	American Pulverizer

DETAILED PROCESS DESCRIPTION - A description of each unit operation used in the upgraded wTe process is presented in the following section. The function of each piece of equipment is described.

Primary Size Reduction System - The first step in the upgraded process was a shear shredder. The machine used was the same 100 HP hydraulically driven machine used for Phase I, but re-configured to its original coarse cut configuration using 3" cutters and removing the 5/8" size control screen. The machine was originally designed for applications involving bulky municipal solid waste and has the ability to accept materials containing heavy pieces of tramp metal without damaging itself. The machine was used to pre-process parts such as automotive bucket seats and whole instrument panels which contain higher levels of heavy-gauge metal than the rotary grinder can handle. A shredder of this configuration only reduces materials down to a nominal 46" particle size, requiring further size reduction for complete materials liberation prior to entering the downstream separation equipment.

Primary Ferrous Metals Removal System - Material exiting the primary shredder dropped onto a conveyor belt which passed under an overhead belt magnet. This device, also a type used in solid waste processing applications, was suspended above the belt conveyor. It contains an electromagnet and a slider belt which moves along the surface of the magnet. The unit was arranged such that the direction of travel for the slider belt was perpendicular to the travel of the conveyor below. As material passed under the magnet, ferrous metals were pulled up out of the shredded material stream by the magnet and held against the slider belt. The slider belt carried the ferrous material away from the magnet and the main conveyor where it was dropped into a container.

Secondary Size Reduction System - The rotary grinder was installed as the secondary size reduction step to further reduce incoming materials from the shear shredder to a smaller (nominal 5/8") particle size and to liberate various attached materials (metals, plastics, paper, foam, etc.) from the plastic parts so that the downstream separation equipment could separate the various fractions from each other. Material fed into an infeed hopper, where a hydraulically actuated pivoting ram system pushed the material against a cutting rotor turning at approximately 70 RPM. The rotor's machined surface contained 64 replaceable cutter knives mounted on its surface with bolts. There was also a stationary knife mounted at the bottom of the hopper through which these cutters pass. A removable size-control screen was mounted below the rotor to control the final output size of the material. This arrangement performed the size reduction process through a combination of cutting and punching of the materials. The intermediate transfer conveyor used to feed materials from the shear shredder to the rotary grinder was arranged to be movable to also allow direct feed of whole parts into the machine's hopper, making it the

primary size reduction device for certain materials, as indicated by the dotted line in Figure 2. Most of the Phase II processing trials used this processing configuration.

Secondary Size Control Screen - The vibrating screen used for secondary size control during the Phase I trials was retained in the process. Through the use of the rotary grinder, a much more uniform particle size was obtained in the shredded material. The amount of oversized materials (>5/8") from this step averaged less than 2% over the entire series of Phase II test runs. This eliminated the need to re-feed materials through the size reduction system, reducing labor and time associated with this extra material handling. As in Phase I, the 1/8" lower deck of the screen was used to remove fines from the process before the material was fed to the downstream separation equipment.

Secondary Ferrous Metals Removal System - A standard drum magnet was used to separate the ferrous materials from the processed materials. This magnet provided a second stage of separation for the small volume of materials pre-processed through the shear shredder, but for the majority of the Phase II materials, it provided the primary separation step. Material was fed onto the surface of a rotating non-magnetic drum. Magnetic materials cling to the drum as its surface passes over an internal fixed magnet. After passing the magnet, the ferrous materials dropped from the drum surface into a discharge chute which deposited the material into a collection bin. The non-ferrous fraction was fed pneumatically into a surge bin with a vibratory feeder used to meter the materials into the air classification system.

Air Classification System - A system employing counter-current airflow was used to remove liberated paper, foam, film, and additional fines from the shredded material prior to feeding it to the wet density separation process. This step was employed to reduce interference with the rigid plastics of extraneous fluff in the sink/float process. The fluff can be more difficult to wet and tends to build up on the surface of the bath.

Density Separation System - As in Phase I, the density separation system consisted of two spiral classifiers, arranged in series, each containing a liquid bath used to separate materials by flotation due to density differences. For Phase II, the primary classifier was modified to provide a greater bath surface area, and the secondary classifier was replaced with a new unit having similar dimensions to the primary unit. Both systems were fitted with spray nozzles and liquid circulation systems to provide improved material handling for the surface materials, and discharge troughs for the light fraction designed to provide additional separation of heavy particles that may be carried over.

Using the different densities of water and salt (calcium nitrate), the specific gravity (ratio of measured fluid density to water density) of the separation media used in the two classifiers was pre-set to allow different materials to float and sink. Material from the size reduction system was fed by gravity from a belt conveyor directly into the first classifier bath, where heavy material would sink and be conveyed by a submerged screw conveyor to the heavies discharge. Light materials, floating to the surface, were conveyed by surface currents to the rear of the vessel where they were discharged in a slurry over the rear edge of the unit to a collection trough and then onto a de-watering screen. The light materials were separated from the liquid by the de-watering screen and were sent either to a mechanical dryer prior to being discharged from the process (primary classifier) or to the next unit operation in the process (secondary classifier). The liquid collected by the de-watering screens was stored in dedicated tanks and pumped back into the appropriate classifier vessel via a circulating pump and spray nozzles. A surfactant was added to both classifier baths to help wet the surface of the incoming materials to aid in the separation process. The capacity of the classifier units was dependent on the volumetric quantity of material floating at each stage. With the increased surface areas of both classifiers, operation was much more stable and required less operator attention. Product throughputs and liquid circulation rates needed to be controlled and matched to allow sufficient residence time for proper separation of the different density fractions while minimizing carryover of the heavy fraction into the lights stream. Since overall classifier throughput capacity is primarily a function of the volumetric flow of floating materials, classifier throughputs were adjusted to compensate for the variations seen in the relative amounts of light and heavy materials in the different product streams being processed. With the exception of fines and lights removal in the size reduction system described above, no further processing to control particle size distribution in the density separation system was performed. No effect on classifier performance due to material particle size distribution was noted during the test runs.

Non Ferrous Metals Removal - The mineral jig, located between the classifiers, was used as an additional metals removal step to eliminate mixed non-ferrous metals, including wiring, molded-in metal fasteners, or non-ferrous metal parts from the product stream. The material entering this unit is all heavier than water. It uses a fresh water bath and a mechanically induced vertical oscillating current to separate the heavier metal components from lighter plastics, capturing the heavier metal fraction in the bottom of a collection basket while allowing the relatively lighter plastics to work their way to the top of the material bed to be discharged into the secondary classifier for further separation of plastics. A spin dryer was used to remove as much of the surface water as possible from the chips before they entered the second classifier in order to minimize density drift over time due to dilution. The metals that accumulated in the collection basket were removed as required on a batch basis.

Washing System - The same paddle mixer from Phase I was used to clean the plastic material that was targeted as product. This unit contained its own water circulation loop to heat the wash water via an indirect heating system. Plastic material was fed in a slurry to a final dewatering screen to recover the wash water and feed the de-watered plastic to a final rinse step and mechanical drying step. Material collected from this drying step was collected as final product for evaluation. The washing system was installed to clean media residue from the target resin stream, as well as oils and grease anticipated to be present on some of the automotive materials. In addition, the system provided some means for removing labels, adhesives, and other coatings, but was not specifically designed for this purpose.

Water Treatment - Fresh water used in the mineral jig, the final rinse step, all spin dryer feeds, and various water sluices for feed assistance were collected in a central process water tank and pumped to the various points in the process requiring fresh water. During operation, the system was operated in a closed loop fashion. At the end of each test run, this water was batch discharged to a separate holding tank to prevent possible contamination between tests. The accumulated process water was disposed of at a wastewater treatment plant.

For the Phase II process runs, a second 60 mesh screen deck was added to the de-watering screens for the classifiers and final wash system to remove additional fine materials from the wet line and to prevent them from building up in the process equipment during operation. For some materials, generation of fines during mechanical agitation associated with the separation equipment was observed. The fine materials collected from the wet system for each test run were weighed and separately identified in the material balances for the Phase II test runs.

Auxiliary Size Reduction Systems - In addition to the in-line size reduction system, two auxiliary pieces of equipment, a 60 HP Hammermill and a 30 HP granulator were installed as stand-alone systems. These were used for specialized testing applications (hammermill) and final size reduction of selected plastic products.

PHASE II AUTOMOTIVE INTERIOR TRIM PROCESSING

DESCRIPTION OF MATERIALS PROCESSED - In contrast to Phase I material collection, during the field collection of materials for the Phase II trials, crews were directed to reject certain parts, such as chrome plated plastics and heavily covered (carpeting, foam/vinyl) pieces as much as possible. In spite of this directive, some heavily contaminated material was received, but the extent of this problem was not discovered until materials were being retrieved from storage for processing. As a result, during the Phase II processing trials, MPRF operating technicians rejected significant amounts of material as it was being fed to the process. This is reported as manual rejects in the material balance summaries.

For the pre-sorted PP and ABS trials, the resin identification system was used to pre-sort the interior trim pieces by resin type. The machine has some capability to differentiate between filled and non-filled PP. Only materials identified as ABS or non-filled PP were segregated as pre-sorted trim for these trials. Heavily contaminated parts were rejected during the pre-sort operation.

Larger quantities of automotive materials were selected to allow sufficiently long processing times to insure stable operation of the system and to eliminate any suspected

Table 3 - Phase II Materials Collected for Processing

Part Description	Target Resin	Qty (lbs)
Automotive Interior Trim	ABS/PP	53679
Automotive Instrument Panels	SMA/ABS	17110
Automotive Seats	Polyurethane Foam	77223
Automotive Bumper Fascias	TPO/PU	10614
Automotive Radiator Endtanks	PA	19148

influence of unstable operation on the Phase II test results. This also provided an opportunity assess other operating logistics (material handling, storage, etc.) at a scale that approximated that of a commercial operation. All of the selected automotive materials actually collected during Phase II are shown in Table 3 below. Over 50,000 pounds of automotive interior trim plastics were collected for the Phase II studies.

As in Phase I, the predominant characteristic of the automotive materials processed was that they consisted of bulky (greater that 12" in at least two dimensions) plastic parts, with high levels of metal contamination in the form of fasteners, structural parts, or other mechanical components. Paints, coatings, and flexible coverings were also present.

TEST PROTOCOL FOR CONTINUOUS PROCESSING
The test protocol for all the Phase II materials was similar to that developed for Phase I. Segregated material groups were continuously processed through the appropriately configured MPRF process line to generate complete material balances, collect material samples, and measure throughput rates, utility consumption, ease of processing, recycle rates, target resin yield, plastics separation performance and labor consumption for each stream. Qualitative observations were also made to compare system performance with Phase I to evaluate and document the process improvements gained. The same sample collection procedure developed for Phase I was used during the Phase II tests. As in Phase I, all the precision density distribution measurements to determine

the percentage of non-target resin reporting to the target density range and the physical properties testing performed for the program were conducted or coordinated by MBA Polymers. Samples were also characterized by resin type using MBA's plastics identification equipment. Selected product samples were sent to APC member companies to perform physical properties measurements.

Additional Phase II Testing Activities - The work completed as part of the Phase II testing included several groups of test runs intended to evaluate the performance of specific unit operations and system modifications prior to initiating the continuous processing trials. These activities included the following:

- A series of benchmarking trials conducted to directly evaluate the performance improvements of the upgraded density separation line using materials re-mixed from Phase I
- A series of preliminary trials conducted to directly evaluate the performance improvements of the upgraded size reduction line processing rigid plastics, focusing on the rotary grinder

Phase II Processing Configuration - The Phase II automotive trim processing trials took advantage of the upgraded size reduction system's improved throughput. This allowed the Phase II trials to be conducted on a continuous basis through the full process line. Processing improvements were noted in terms of overall throughput, product uniformity, and operating labor requirements.

wTe worked with MBA to develop an operations test protocol designed to maximize the quality of the recovered resin streams, minimizing the effect of non-plastic contaminants, which were found to be present in spite of efforts to reject them during the collection process. The protocol called for the manual removal of front-end rejects (heavily covered plastic pieces, pressboard, chrome plated parts, etc.) and for analyzing grab samples of chips from the product streams using the resin identification system during operation to more closely monitor the system. In all cases, the final product was granulated and re-air classified to remove additional contaminants and to produce a product more consistent with commercially available recycled plastic flake.

For the mixed trim runs, the system was set up to make two-passes to recover PP and ABS product streams from these streams. For the first pass, the primary classifier was run dry and the secondary classifier was run with water to separate and wash all plastic having a density less than 1.0 g/cc as a PP product. For the second pass, the system was returned to its normal two-classifier configuration with densities of 1.03 and 1.09 g/cc to target resins between those densities as an ABS product. The material which had been collected as CII heavies (density greater than 1.0 g/cc) during the first pass was used as the feedstock for this run. This was done to remove filled

PP contaminants as floaters in the first classifier, and to remove heavy non-targeted plastics such as PVC, filled nylon, as heavies, and an ABS product stream as floaters. For the presegregated material streams, the first pass configuration described above was used to recover the PP, and the second pass configuration was used for the ABS.

Processing Trial Objectives - The objective of both the Phase I and Phase II trials was to produce cleaned streams of PP and ABS from automotive interior trim obtained from auto dismantlers. In both cases, mixed (non-sorted) and pre-segregated streams (PP and ABS) were processed during a series of runs. This approach allowed an evaluation of the effect of pre-sorting on product quality to be made.

It was anticipated that it would be difficult to produce a pure ABS product stream from an unsorted trim stream with the density separation system alone due to the presence of filled PP in some of the parts, because of the density overlap. In the Phase II trials, additional care was taken to eliminate items such as chrome-plated plastics and heavily covered materials from the infeed stream in order to produce the highest quality product possible under a best-case scenario.

Seven processing trials on automotive trim were completed at the MPRF as part of the Phase II test series. Table 4 provides a summary description for each.

Table 4 - Interior Trim Processing

wTe Test ID	Source Material	Test Description	Objective
M-131 1B	Mixed Interior Trim	Highly contaminated parts manually rejected during process feed; Two pass configuration used	Recover maximum quality PP and ABS product with mechanical processing only
M-131 2B	Pre-Sorted PP	Pre-sorted material processed with single pass configuration for PP; Minimal additional decontamination	Recover maximum quality PP product. Demonstrate effect of MIR technology on product quality
M-131 3B	Pre-Sorted ABS	Pre-sorted material processed with single pass configuration for ABS; Minimal additional decontamination	Recover maximum quality ABS product; Demonstrate effect of resin identification technology on product quality
M-131 4B	Front-End Rejects from Tests 1B-3B	Rotary grinder, screen, and air classifier only	Preliminary worst-case assessment of size reduction system's ability to handle high contamination levels
M-131 6B	Mixed Interior Trim	Contaminated trim parts not rejected during process feed – only non-trim; Two pass configuration used	Evaluate system capabilities with as received materials; For comparison to test M131-1B
M-131 7B	Mixed Interior Trim	Contaminated trim parts not rejected during process feed – only non-trim; Two pass configurations used	Evaluate system capabilities with as received materials; for comparison to tests M131-1B & 6B
M-131 14B	Mixed Interior Trim	Highly contaminated trim and non-trim parts rejected during process feed; Two pass configuration used	Evaluate large scale processing system capabilities with remaining material. Compare results with other mixed trim tests.

Special Considerations-Automotive Interior Trim - The final processing trial (Test M131-14B) in the Phase II series represented the largest of the processing trials conducted at wTe as part of the APC program. A truckload of material was processed through the upgraded mechanical processing line to confirm performance data collected on smaller trials, as well as to evaluate the operating economics of the process on a reasonably large scale.

Comparison of Automotive Interior Trim Processing Options - During processing of automotive interior trim at the MPRF, three levels of material processing were employed. Table 5 summarizes the three approaches, and lists advantages and disadvantages, as they relate to the then current MPRF process line, for each one.

efficiently recovered for processing in the MPRF line. A higher than anticipated front-end reject percentage (approximately 45% vs a high of about 25% seen previously) was measured during processing of the material, primarily because the pile had already been "mined" of the higher quality (less contaminated) materials during the previous trials.

In order to compensate for some of these effects, the contaminated trim was pre cleaned or rejected at the storage bunker as it was manually retrieved for processing inside the building. The extra material handling and pre-screening work required at least one, and sometimes two additional people dedicated to that activity. The labor associated with this operation was not included in the

Table 5 – Automotive Interior Trim Processing Options

Processing Description	Advantages	Disadvantages
Level I Minimal Contaminant Rejection; No pre-sorting w/resin ID equipment	Lowest operating cost	Lower process throughputs due to high fluff content of parts; Higher risk of lower product quality
Level II Aggressive Contaminant Rejection; No pre-sorting w/resin ID equipment	Small incremental operating cost increase; less material related equipment problems; increased product quality likely; Can produce a usable product, especially PP. Could be enhanced by lowering blend ratio	ABS product quality likely to be marginal
Level III Aggressive Contaminant Rejection; w/pre-sorting Using resin ID equipment	Produces highest quality Products	Highest cost due to labor intensive pre-sort; No significant quality benefit for PP fraction

The automotive interior trim collected by wTe was processed in two steps (passes) to recover a polypropylene and an ABS product. This material had been stored outdoors in storage bunkers located at the MPRF. It had been moved during the course of the program with a small skid-steer loader tractor which resulted in additional breakage into small pieces. The pile had also been picked through during the course of conducting prior trials for the APC program. In particular, approximately 7,500 lbs of mixed trim had been selectively hand picked and pre-sorting in order to provide feedstock samples of non-filled PP and ABS trim for test runs M131-2B and M131-3B. As a result of these activities, the composition of the remaining pile was believed to contain a higher content of filled polypropylene, overly contaminated parts, and non target resins, as well as small pieces and chips which were not able to be

economic evaluations for the process. The selected pieces were loaded into 2.2 cu yd dump hoppers which were then transported into the building. Each dump hopper held an average of 180 pounds of plastic parts, and in all, eighty two hopper trips were required to transfer the material into the building.

As had been done during the previous mixed trim runs, the heavy material fraction collected during the first pass, which was performed to recover the polypropylene product, was re-run through the normal dual-classifier configuration to separate as much of the filled PP from the ABS as possible. Before running the second pass of test M131-14B to recover ABS, wTe personnel reviewed the data from the earlier Phase II mixed trim tests (M131-1B and 7B). For test M131-1B, the density of the second classifier had been set at 1.09 g/cc. Due to a density drift discovered during the run, test M131-7B, had been set at

1.08 g/cc. In comparing the results, use of the higher density bath in the second classifier had resulted in higher measured ABS product purities (90.6% for the 1.09 g/cc run vs. 83.3% at 1.08 g/cc). It also resulted in lower ABS losses to the CII heavy fraction (15% vs. 43%) for the two runs. Because the measured levels of filled PP were similar for the two runs, (5.8% and 6.7% respectively) it appeared that the most significant effect of the narrower density cut was to increase the amount of target ABS that was being lost in CII.

Based on this information, bath densities of 1.03 g/cc and 1.09 g/cc were selected for the final interior trim run in order to maximize product yield. During the run, however, grab samples of the target resin (ABS) product were analyzed and higher levels of filled PP (spikes above 10%) were observed to be reporting to the product. It was hypothesized that the suspected higher concentration filled PP in the infeed stream was contributing to additional carryover, and the bath density in CII was lowered from 1.09 to 1.085 g/cc to compensate. This resulted in a more pure output product but increased losses of ABS to the CII heavy fraction. Product characterization using the ID equipment was continued on a spot-check basis for all except the last hour (estimate) of the run due to an equipment malfunction. The resulting product purity reported from the available data was 75.9% ABS content with a 10.6% filled PP content.

Table 6 shows the detailed Phase II processing and mass balance results for automotive interior trim.

DISCUSSION OF RESULTS – AUTOMOTIVE TRIM

The data obtained from the automotive interior trim Phase I and Phase II processing trials (Tables 1 and 6) shows that significant performance improvements were gained through the implementation of the process upgrades made to the MPRF process line. These improvements can be summarized as:

- The upgraded MPRF process is capable of continuous processing of contaminated interior trim parts through the full process line.
- Size reduction system throughput increased from 238 lbs/hr to 778 lbs/hr
- Density separation system throughput was increased from 277 lbs/hr to 671 lbs/hr (based on presorted PP runs)
- The level of non-target resin contaminants in the PP product was reduced to less than 1% in both mixed trim and pre-sorted PP trim processing trials
- The levels of non-target resin contaminants in the ABS product were reduced to less than 13% in the mixed trim processing trials; These levels are attributed to the presence of filled PP
- The level of non-target resin contaminants in the ABS product was reduced to less than 1% in the pre-sorted ABS trim processing trials
- The difference between the contamination levels seen in the non-sorted vs. the pre-sorted ABS

demonstrates the benefit of employing resin identification technology where density overlap exists in the parts to be processed.

For comparison, both the Phase I and Phase II data are presented side-by-side in Tables 7 and 8. Table 7 presents a consolidation of the data for the mixed trim trials; Table 8 presents similar data for the pre-sorted trim trials.

In all, approximately 27,500 lbs of material was recovered as cleaned product from 40,000 lbs of automotive trim processed. Samples of selected product materials were sent to APC member companies for physical properties testing to evaluate product quality. Physical properties were measured and compared against those of selected control materials. In general, the measured properties exhibited lower values than those of the control samples, particularly in the area of low temperature impact strength. In all cases, the properties exhibited by the samples produced from the pre-segregated material streams were better than those from the mixed PP & ABS product streams.

SEPARATION OF FILLED PP AND ABS

As noted above, several weight percent of filled PP appeared as a contaminate in the recovered ABS due to density overlap. It was obvious that some method other than density separation would be required to remove the filled PP from the ABS. APC approached Recovery Plastics International (RPI) in Salt Lake City, Utah to evaluate their "skin flotation" technology to affect this separation. Using hand sorting in conjunction with mid-infrared ID, color-coded samples of end-of-life vehicle ABS and filled PP were prepared by wTe and MBA Polymers and sent as a mix to RPI. RPI successfully achieved separation of these two plastics using a non-density based flotation process.

PROCESS ECONOMICS

A preliminary assessment of MPRF operating economics was made using a simple cost model. The model utilized the capital costs associated with the MPRF process and assumed continuous processing. In the case of mixed automotive interior trim, the model incorporated a third sink-float classifier to avoid the two-pass processing employed during Phase II testing. Variable operating costs include all labor, utilities, residue disposal, and maintenance. Overhead costs such as facility rent, management labor, and debt service on equipment are treated as a separate line item from the variable operating costs. The model generates unit costs as $/lb of infeed material based on a single shift, 20 day/month operating basis. The model does not incorporate the cost associated with collecting the plastics at a dismantling facility and shipping them to the plastics processing facility. Based on Phase II processing of interior trim plastics, unit costs were calculated to be as high as $0.90/lb. It was obvious that low material throughput was a major contributor to the high costs. To achieve a $.20-

$.30 range for unit processing cost, the model demonstrated that material input at the MPRF must be above 2000 pounds per hour. This would necessitate making modifications to both the size reduction and density separation equipment.

CONCLUSION

A sufficiently large quantity (over 50,000 pounds in total) of plastic interior trim parts from end-of-life vehicles have been processed using a combination of dry and wet methods to benchmark mechanical recycling potential.

Process capability has been documented for various interior trim feedstock combinations showing the ability to significantly upgrade both unfilled PP and ABS resins from less than 50% to greater than 90% purity. Due to the presence of filled PP in the ABS product, further purification of the ABS should employing non-density based separation methods.

The utility of rapid manual mid-infrared spectroscopy to presort ELV interior trim parts by resin type prior to processing has been demonstrated and quantified but throughputs are low and resulting economics are marginal at best. ID technology has been demonstrated to have considerable utility in assisting with in-process and final product quality control.

A preliminary economic model for the plastic processing steps at the MPRF suggest that processing costs of about $0.25 per pound of final product should be accessible at a throughput of 2000 pounds per hour of pre-sorted interior trim.

ACKNOWLEDGEMENTS

This work would not have been possible without the vision and support of APC member companies. The authors would like to especially thank the Vehicle Recycling Partnership and the Automotive Recyclers Association for their assistance during the parts collection phase of the study.

CONTACT

For further information, please contact:

Dr. Michael A. Fisher
Director
American Plastics Council
1300 Wilson Blvd.
Suite 800
Arlington, VA 22209
Phone (703) 253-0614, or FAX (703) 253-0701
Email: Mike_Fisher@plastics.org.

Table 6 – Phase II Test Results for Automotive Interior Trim

Test Report No:	M 131 1B	M 131 2B	M 131 3B	M 131 4B
Target Plastic:	PP&ABS	PP	ABS	PP&ABS
Description:	Mixed Trim	PP Trim	ABS Trim	Mixed Trim
Size Reduction System Material Balance				
Infeed Material (to primary size reduction)	9,066	2,350	2,746	2,326
Rejects - Manual Inspection	2898	72	22	196
Ferrous Metal (primary)	0	0	0	0
Oversized Screened Material	36	134	82	100
Undersized Screened Material	384	158	162	174
Ferrous Metal (secondary)	128	48	74	234
Air Classifier Lights	122	104	66	506
System Cleanup - Size Reduction Line	66	74	54	86
Wet Process System Material Balance				
Infeed to Wet Process	5,432	1760	2,286	1,030
Primary Classifier Lights	246	0	134	0
Mineral Jig Heavy Fraction	76	20	18	
Secondary Classifier Heavy Fraction	240	38	46	
Water Treatment System De-watering Screen Solids	12	18	n/a	
Wash System De-watering Screen Solids	5	2	n/a	
System Cleanup - Separation/Wash Line	252	78	148	
Q/C Sampling	104	20	30	
System Losses (by difference)	115	(30)	(6)	16
Target Plastic Product (1st pass - mixed trim runs)	3,010	1614	1916	1014
Target Plastic Product (2nd pass - mixed trim runs)	1,372	-	-	-
Target Plastic Product (total - includes Q/C samples	**4,486**	**1,634**	**1,946**	**1,014**
Process Yields				
Size Reduction System Yield (lbs into wet process/lbs into size reduction-incl rejects)	60%	75%	83%	44%
Wet Process Sytem Yield (targe resin out/lbs into wet process)	83%	93%	85%	98%
Overall Yield – Target Plastic (lbs target resin/lbs into size reduction – incl. Rejects)	49%	70%	71%	44%
Adjusted Yield – Target Plastic (lbs target resin/lbs into size reduction – excl. rejects)	73%	72%	71%	48%
Production Summary				
Process Material Throughput	778	671	392	1,642
Recycle Rate (Lbs Re-run/Lbs In)	0%	0%	0%	0%
Processing Hours (Mat'l Feed Time)	11.57	3.50	7.00	1.42
Total Hours (Including Setup & Downtime)	16.58	3.50	7.66	1.42
Utilization Factor (Processing/Total Hrs)	70%	100%	91%	100%
Classifier Bath Densities				
First Pass:				
C1 Bath Density (g/cc)	DRY	DRY	1.03	DRY
C2 Bath Density (g/cc)	1.00	1.00	1.09	1.00
Second Pass:				
C1 Bath Density (g/cc)	1.03			1.03
C2 Bath Density (g/cc)	1.09			1.09
Summary				
Yield (% of whole parts in less manual rejects)				
Target Plastic	72.7%	71.7%	71.4%	47.6%
Ferrous metals	2.1%	2.1%	2.7%	11.0%
Non Ferrous Metals	1.2%	0.9%	0.7%	0.0%
Other	24.0%	25.3%	25.2%	41.4%

Table 6 – Phase II Test Results for Automotive Interior Trim (continued)

Test Report No: Target Plastic: Description:	M 131 6B PP&ABS Mixed Trim	M 131 7B PP&ABS Mixed Trim	M 131 14B PP&ABS Mixed Trim
Size Reduction System Material Balance			
Infeed Material (to primary size reduction)	5,212	5,060	26,702
Rejects - Manual Inspection	334	728	12,064
Ferrous Metal (primary)	0	0	0
Oversized Screened Material	26	12	30
Undersized Screened Material	318	260	820
Ferrous Metal (secondary)	256	158	374
Air Classifier Lights	374	184	266
System Cleanup - Size Reduction Line	140	96	116
Wet Process System Material Balance			
Infeed to Wet Process	3,764	3622	13,032
Primary Classifier Lights	50	74	202
Mineral Jig Heavy Fraction	76	48	0
Secondary Classifier Heavy Fraction	316	206	756
Water Treatment System De-watering Screen Solids	18	22	26
Wash System De-watering Screen Solids	20	10	22
System Cleanup - Separation/Wash Line	192	222	590
Q/C Sampling	44	52	196
System Losses (by difference)	96	(16)	(632)
Target Plastic Product (1st pass - mixed trim runs)	2,180	2,358	8,644
Target Plastic Product (2nd pass - mixed trim runs)	772	646	3,228
Target Plastic Product (total - includes Q/C samples	**2,996**	**3,056**	**12,068**
Process Yields			
Size Reduction System Yield (lbs into wet process/lbs into size reduction -incl. rejects)	72%	72%	45%
Wet Process System Yield (target resin out/lbs into wet process)	80%	84%	93%
Overall Yield - Target Plastic (lbs target resin/lbs into size reduction - incl. rejects)	57%	60%	45%
Adjusted Yield - Target Plastic (lbs target resin/lbs into size reduction - excl. rejects)	61%	71%	82%
Production Summary			
Process Material Throughput	711	578	592
Recycle Rate (Lbs Re-run/Lbs In)	0%	0%	0%
Processing Hours (Mat'l Feed Time)	7.33	8.75	45.1
Total Hours (Including Setup & Downtime)	10.66	10.42	64.0
Utilization Factor (Processing/Total Hrs)	69%	84%	70%
Classifier Bath Densities			
First Pass:			
C1 Bath Density (g/cc)	DRY	DRY	DRY
C2 Bath Density (g/cc)	1.00	1.00	1.00
Second Pass:			
C1 Bath Density (g/cc)	1.03	1.03	1.03
C2 Bath Density (g/cc)	1.09	1.08	1.085
Summary			
Yield (% of whole parts in less manual rejects)			
Target Plastic	61.4%	70.5%	82.4%
Ferrous metals	5.2%	3.6%	2.6%
Non Ferrous Metals	1.6%	1.1%	0.0%
Other	31.8%	24.7%	15.0%

Table 7 - Consolidated Summary – Mixed Interior Trim Runs

	M 131-2	M 131 1B	M 131 6B
Test Report No:			
Target Plastic:	PP&ABS	PP&ABS	PP&ABS
Description:	Mixed Trim	Mixed Trim	Mixed Trim
Size Reduction System Material Balance			
Infeed Material (to primary size reduction)	11,138	9,066	5,212
Rejects - Manual Inspection		2898	334
Ferrous Metal (primary)		0	0
Oversized Screened Material	0	36	26
Undersized Screened Material		384	318
Ferrous Metal (secondary)	90	128	256
Air Classifier Lights	528	122	374
System Cleanup - Size Reduction Line	800	66	140
Wet Process System Material Balance			
Infeed to Wet Process	9,720	5432	3764
Primary Classifier Lights	5,076	246	50
Mineral Jig Heavy Fraction	368	76	76
Secondary Classifier Heavy Fraction	746	240	316
Water Treatment System De-watering Screen Solids		12	18
Wash System De-watering Screen Solids		5	20
System Cleanup - Separation/Wash Line		252	192
Q/C Sampling	240	104	44
System Losses (by difference)	80	115	96
Target Plastic Product (1st pass - mixed trim runs)	5076	3,010	2,180
Target Plastic Product (2nd pass - mixed trim runs)	3210	1,372	772
Target Plastic Product (total - includes Q/C samples	**8,526**	**4,486**	**2,996**
Production Summary			
Size Reduction System Yield (lbs into wet process/lbs into size reduction -incl. rejects)	**87%**	**60%**	**72%**
Wet Process System Yield (target resin out/lbs into wet process)	**88%**	**83%**	**80%**
Overall Yield - Target Plastic (lbs target resin/lbs into size reduction - incl. rejects)	**77%**	**49%**	**57%**
Adjusted Yield - Target Plastic (lbs target resin/lbs into size reduction - excl. rejects)	**77%**	**73%**	**61%**

Table 7 - Consolidated Summary – Mixed Interior Trim Runs (continued)

	M 131-2	M 131 1B	M 131 6B
Test Report No:			
Target Plastic:	PP&ABS	PP&ABS	PP&ABS
Description:	Mixed Trim	Mixed Trim	Mixed Trim
Size Reduction Process			
Process Material Throughput	238	778	711
Recycle Rate (Lbs Re-run/Lbs In)	51%	0%	0%
Processing Hours (Mat'l Feed Time)	46.8	11.57	7.33
Total Hours (Including Setup & Downtime)	72.0	16.58	10.66
Utilization Factor (Processing/Total Hrs)	65%	70%	69%
Wet Separation Process			
Process Material Throughput	508		
Processing Hours (Mat'l Feed Time)	20.2		
Total Hours (Including Setup & Downtime)	40.0		
Utilization Factor (Processing/Total Hrs)	50%		
Classifier Bath Densities:			
First Pass:			
C1 Bath Density (g/cc)	1.00	DRY	DRY
C2 Bath Density (g/cc)	1.08	1.00	1.00
Second Pass:			
C1 Bath Density (g/cc)		1.03	1.03
C2 Bath Density (g/cc)		1.09	1.09
Summary			
Yield (% of whole parts in less manual rejects)			
Target Plastic	76.5%	72.7%	61.4%
Ferrous metals	0.8%	2.1%	5.2%
Non Ferrous Metals	3.3%	1.2%	1.6%
Other	19.3%	24.0%	31.8%

Table 8 - Consolidated Summary – Pre-Sorted Interior Trim Runs

	M 131-3	M 131 2B	M 131-4	M 131 3B
Test Report No:				
Target Plastic:	PP	PP	ABS	ABS
Description:	PP Trim Pre-Sorted	PP Trim Pre-Sorted	ABS Trim Pre-Sorted	ABS Trim Pre-Sorted
Size Reduction System Material Balance				
Infeed Material (to primary size reduction)	7,160	2,350	1,968	2,746
Rejects - Manual Inspection		72		22
Ferrous Metal (primary)		0		0
Oversized Screened Material	0	134	0	82
Undersized Screened Material		158		162
Ferrous Metal (secondary)	64	48	20	74
Air Classifier Lights	398	104	32	66
System Cleanup - Size Reduction Line	338	74	112	54
Wet Process System Material Balance				
Infeed to Wet Process	6,360	1,760	1,804	2,286
Primary Classifier Lights	n/a	0	102	134
Mineral Jig Heavy Fraction	4	20	16	18
Secondary Classifier Heavy Fraction	134	38	82	46
Water Treatment System De-watering Screen Solids		18		n/a
Wash System De-watering Screen Solids		2		n/a
System Cleanup - Separation/Wash Line		78		148
Q/C Sampling	92	20	24	30
System Losses (by difference)	998	(30)	102	(6)
Target Plastic Product (1st pass – mixed trim runs)	5,132	1614	1,478	1916
Target Plastic Product (2nd pass – mixed trim runs)		-		-
Target Plastic Product (total – includes Q/C samples	5,224	1,634	1,502	1,946
Production Summary				
Size Reduction System Yield (lbs into wet process/lbs into size reduction -incl. rejects)	**89%**	**75%**	**92%**	**83%**
Wet Process System Yield (target resin out/lbs into wet process)	**82%**	**93%**	**83%**	**83%**
Overall Yield - Target Plastic (lbs target resin/lbs into size reduction - incl. rejects)	**73%**	**70%**	**76%**	**71%**
Adjusted Yield - Target Plastic (lbs target resin/lbs into size reduction - excl. rejects)	**73%**	**72%**	**76%**	**71%**

APPENDIX

AUTOMOTIVE INTERIOR TRIM PRESORT OPERATIONS USING MANUAL MIR EQUIPMENT Approximately 10,000 lbs. of the ABS and PP automotive interior trim from end-of-life vehicles was identified and pre-sorted in this study using a Bruker rapid plastics mid infrared instrument (P/ID 28).

Pre-sorting using the MIR instrument proved to be extremely time consuming. In an effort to gain a more detailed understanding of the material handling aspect of this process, time study data was recorded to help quantify the methods employed. This information was used to compare the results with reported results from similar sortation operations in order to assess the progress of the sorting operation. The time study data presented in this section was collected early in the pre-sorting process.

Procedure - Before the 10,000 lb. sorting operation was initiated, a series of time studies were conducted to study the effect of material handling methods on pre-sort production rates. The mixed interior trim was retrieved from the storage piles at the MPRF using a Bobcat loader and the 2.5 cu.yd. dump hoppers. Material was then sorted by placing individual trim pieces in front of the P/ID 28's detector head, obtaining a reading, and depositing the piece in the appropriate gaylord box depending on the reading obtained. Material was sorted into fractions identified as "PP" (including "PP-EPDM"), "ABS", "PP-w/filler" (10%, 20%, 40% talc or 40% chalk), or "Other" (SB, HIPS, SMA, PC, PC+ABS, PPE+SB (PPO), ASA, and SMC. After approximately 2,800 lbs of material had been sorted, average piece weights were obtained for each of these fractions. Because the MIR equipment uses surface analysis, some of the pieces required that surface grinding be employed to clean off paint or contamination that prevented an accurate reading of the actual resin. The time study data also collected measured the extent to which cleaning and re-analysis was required. The information was used to project and schedule manpower for the final sortation operation.

Characterization of Sorted Material - The breakdown of sorted interior trim by resin type is reported below in Table 9, along with the piece weight data. The front end reject fraction represented trim pieces, such as door panels (plastic and pressboard), which were extensively covered with vinyl skins, carpeting, or a combination of both. In the case of plastic substrates, removal of these coverings was attempted but some were so well adhered that removal was not practical and the entire piece was rejected. Interior trim pieces with chrome plated coatings were also rejected. The reject stream also included non-plastics such as sound deadening material, carpeting, speakers, and other metal components which were removed from the acceptable trim pieces.

Table 9 – Interior Trim Characterization - Interim Check

Identification	Pounds	Percent	Avg #/Pc.	# Pieces
PP/EPDM	1,536	55%	1.17	1,313
ABS	514	19%	0.93	553
PP w/filler	86	3%	0.99	87
Front end reject	186	7%	N/a	N/a
Other resins	98	4%	1.20	82
Trash	352	12%	N/a	N/a
Total	2,772	100%		

Interim Time Study - Identification Step - Table 10 summarizes the measured times to perform each task associated with the piece identification step of the pre-sorting process. "Analysis Time" represents the period from when the Bruker pedal is actuated until the monitor displays its findings. "Analyze and box" represents the time period from when the pedal is actuated, identification is displayed, and the piece is put into its respective box which is less than eight feet away. The rest of the task descriptions represent the different series of motions performed to obtain the part identification, depending on various situations encountered as shown. These times include picking up a piece of trim (P/U) from a sample pile located less than eight feet away, performing the analysis, and placing the piece into a designated box located less than eight feet away.

Table 10 – Time Study of Automotive Pre-sort Tasks

Task Description	Seconds
Analysis time	5
Analyze and box	8
P/U, analyze, and box	11
P/U, grind, analyze, and box	18
P/U, analyze, grind, re-analyze, and box	25
P/U, analyze, label, and box	20
P/U, grind, analyze, label and box	31
P/U, analyze, grind, re-analyze label, and box	44

Impact of Surface Contamination on Sample Readability and Production Rate - During the sortation time study, the operators noted that the need for surface grinding using a hand-held electric grinder differed for the "PP", "ABS", "PP-w/filler", and "Other" resin categories. Surface preparation was considered necessary when readings such as "PMMA" (generally indicating paint) or "Hit rate too high" (indicating that the Bruker system could not obtain a suitable match between the reading and the resin spectra library) were obtained. For both PP categories, only 10% of the pieces analyzed required grinding, as compared to 50% of the ABS and 100% of the "Other" categories. It has been speculated that in the case of the ABS, at least, the erroneous readings obtained prior to grinding could be due to surface oxidation effects, particularly where "SAN" readings were obtained on ABS samples prior to grinding.

Using the data presented in Tables 9 and 10, a theoretical rate for the piece identification step was calculated for each resin type and is presented in Table 11. These figures represent the MIR identification step only (part pickup, grinding, and boxing), and don't include the "overhead" time spent on material retrieval from storage, part cleaning, sample labeling, and data collection. The total time to sort the 2,772 lbs of trim shown in Table 9 was 56 hours, yielding an overall processing rate (including "overhead" activities) of approximately 50 lbs/hr. (based on contaminated material infeed) for the material pre-segregation operation. The differences in calculated sort time for the different resin types were due primarily to the different unit piece weights and different percentages of pieces requiring grinding and removal of trash and reject material. The large difference in the actual vs. theoretical production rates (50 vs. 260 lbs/hr) indicated that the critical objective, pre-sorting of 3,000 lbs of ABS, could best be improved by concentrating on the peripheral material handling areas.

Table 11 - Theoretical Automotive Trim Pre-sort Rates

Identification	Pounds	Sort time	#/hour
PP	1,536	4.5	339.8
ABS	514	2.8	186.2
PP w/ filler	86	0.3	286.7
Other resins	98	1.0	98.0
Total	2,234	8.58	260.4

Comparison With Other Reported Data - BMW has reported ("Rapid Identification of Plastics" by Carl Hanser Verlag, Kunststoffe, May, 1994) production rates in a similar sorting operation using Bruker MIR equipment of approximately 240 lbs/hr (all parts) based on analysis times of 20 seconds per part and an average part weight of 1.32 lbs.   This compares with an overall sorting production rate of 260 lbs/hr, an analysis time, including surface preparation, of 18 seconds, and an average part weight of 1.10 lbs, in the present study.

Modified Pre-sort Procedure – In order to expedite the pre-sorting effort, the procedure was revised to incorporate a selective ABS removal procedure from the storage pile. Since the rate at which ABS was being collected from the pile and segregated during the random collection operation was very low, the technicians began selectively picking trim from the storage bunker which appeared to be

ABS. ABS appears to be more brittle and sounds different when scratched or impacted than PP.

Additional time study data was collected after implementing this process to assess its effectiveness. A comparative summary of the new, "ABS Selective" collection data, along with data from the "Random" collection method is presented in Table 12. Random collection refers to the non-ABS selective material retrieval the technicians initially followed to collect the interior trim for identification. The data reported in Table 12 was compiled at a later time during the sortation process than when reported previously in Table 9. During that time, the overall recovery rate dropped to approximately 30 lbs/hr from the previously estimated 50 lbs/hr. The earlier reading had been based on a small sample size and could have been less accurate. The material was collected in a 2.5 yard dump hopper and brought into the building for identification. Inside the building the trim was inspected, cleaned, identified, placed into boxes and recorded.

The time study data collected showed that by selectively targeting ABS, the percent by weight of material pre-sorted increased from 19 percent to 74 percent. The corresponding ABS collection rate increased from 5.71 pounds per hour to 26.41 pounds per hour, significantly speeding up the collection pre-sorting process.

USE OF THE BRUKER P/ID 28 IDENTIFICATION SYSTEM FOR ON-LINE SORTING
Sample Collection and Analysis Procedure - The Bruker plastics identification system was also used to monitor the density separation system's performance during selected mixed interior trim processing runs by checking for non-target resins in grab samples of various output streams during operation. Operators evaluated grab samples (30 chips) of the product every 15 minutes (120 chips/hr), and grab samples of the CI and CII streams every hour (30 chips/hr). As an example, the results of this quality monitoring procedure for production of an ABS product from a mixed trim infeed source are summarized in the tables below. Table 13 shows characterization data of output samples collected during the second pass run. The resin types are grouped together as shown in the table into the following groups: "ABS", "non-filled PP", "filled PP", and "Other".

The sample output analysis was done over a four hour operating period; the total number of ABS product chips analyzed was 480 and the total number of chips from the CI and CII classifiers was 120 for each stream. Based on the data shown in Table 13 (first pass) and Table 14 (second pass), the system clearly was able to reduce the non-ABS content of this highly mixed stream from 56.8% to 8.1%. Table 15 summarizes target resin characterizations measured by wTe for selected interior trim runs. This data was obtained by MBA Polymers from composite product samples generated during the runs.

This information can be used to benchmark the starting point for more refined density separation methods or other technical approaches as they become available. As part of future work, further confirmation of the accuracy of the plastics identification system should also be included, although wTe's preliminary testing has shown the unit to have high accuracy when sample surfaces are properly prepared.

Table 12 - ABS Selective Time Study Data

Resin Identification	Random Collection 154 Man Hours Elapsed			ABS Selective 70.5 Man Hours Elapsed			Overall Collection 224.5 Man Hours Elapsed		
	Pounds	Percent	Yield #/hr	Pounds	Percent	Yield #/hr	Pounds	Percent	Yield #/hr
PP	2,330	50%	15.13	60	2%	0.85	2,390	34%	10.65
ABS	880	19%	5.71	1,862	74%	26.41	2,742	38%	12.21
PP w/filler	128	3%	0.83	20	1%	0.28	148	2%	0.66
Front end reject	434	9%	2.82	198	8%	2.81	632	9%	2.82
Other resins	228	5%	1.48	122	5%	1.73	350	5%	1.56
Trash	616	13%	4.00	246	10%	3.49	862	12%	3.84
Total	4,616	100%	29.97	2,508	100%	35.57	7,124	100%	31.73

Table 13 - CII Heavy Fraction Characterization
On-Line Sampling of C II Heavies from First pass of Mixed Trim Run

	Target: ABS #	ABS #	Non-Target #	Filled PP #	Total ABS %	Total Non-ABS %
	120	48	44	28	40.0%	60.0%
Chip						
Count:	120	54	42	24	45.0%	55.0%
	120	72	31	17	60.0%	40.0%
	120	30	56	34	25.0%	75.0%
	120	51	46	23	42.5%	57.5%
	60	30	20	10	50.0%	50.0%
	660	285	239	136	-	-
Contamination	(%)	43.2%	36.2%	20.6%	43.2%	56.8%

Table 14 - On-Line Product Quality Monitoring
Second Pass of MPRF Mixed Trim Processing Trial

Resin ID	Product	C1	C2
ABS	90.6%	40.0%	15.0%
ASA	1.3%	0.8%	0.0%
Total ABS	**91.9%**	**40.8%**	**15.0%**
PP	0.2%	14.2%	15.8%
PP EPDM	2.1%	0.0%	0.0%
Total Non-Filled PP	**2.3%**	**14.2%**	**15.8%**
PP 20%T	0.4%	15.0%	4.2%
PP 40%T	2.1%	12.5%	22.5%
PP 10%T	1.3%	3.3%	0.0%
PP 40%C	0.0%	0.8%	3.3%
Total Filled PP	**3.8%**	**31.7%**	**30.0%**
HIPS	1.0%	10.0%	5.0%
PBT	0.2%	0.0%	0.0%
PC	0.0%	0.0%	11.7%
PC+ABS	0.0%	0.0%	16.7%
PPE+SB	0.8%	3.3%	2.5%
PA66	0.0%	0.0%	2.5%
SMC	0.0%	0.0%	0.8%
Total Other	**2.1%**	**13.3%**	**39.2%**

Table 15 - Selected Product Quality Measurement Results – Automotive Trim

	wTe Trial No. Infeed Material	M 131 6B Mixed Trim 1st Pass	M131 7B Mixed Trim 1st Pass	M 131 14B Mixed Trim 1st Pass	M 131 2B Sorted PP		
	Target Resin:	PP	PP	PP	PP		
ABS		0.0%	0.0%	0.0%	0.0%		
All Styrenics		0.0%	0.0%	0.2%	0.2%		
PP & PP (EPDM)		97.5%	97.9%	97.5%	93.6%		
Filled PP		0.0%	1.4%	1.9%	0.4%		
PUR		0.0%	0.0%	0.0%	n/a		
SMA		0.0%	0.0%	0.0%	n/a		
Other/Nylon		2.5%	0.7%	0.4%	n/a		
	wTe Trial No. Infeed Material	M 131 1B Mixed Trim 2nd Pass	M 131 6B Mixed Trim 2nd Pass	M 131 7B Mixed Trim 2nd Pass	M 131 14B Mixed Trim 2nd Pass	M 131 3B Sorted ABS	M 131 3B Sorted ABS
	Target Resin:	ABS	ABS	ABS	ABS	ABS	ABS
		90.6%	68.8%	8.3%	75.9%	98.5%	99.6%
ABS		90.6%	68.8%	83.3%	75.9%	98.5%	99.6%
All Styrenics		92.9%	77.5%	86.7%	87.6%	99.3%	99.6%
PP & PP (EPDM)		0.2%	3.8%	6.7%	0.0%	0.0%	0.0%
Filled PP		5.8%	16.3%	6.7%	10.6%	0.0%	0.0%
PUR		0.0%	0.0%	0.0%	0.0%	n/a	n/a
SMA		0.0%	0.0%	0.0%	0.0%	n/a	n/a
Other/Nylon		1.0%	2.5%	0.0%	1.8%		
	wTe Trial No. Infeed Material	M 131 2B Sorted PP	M 131 3B Sorted ABS	M 131 3B Sorted ABS	M 131 3B Sorted ABS		
	Target Resin:	PP	ABS	ABS	ABS		
ABS		0.0%	98.5%	99.6%	96.2%		
All Styrenics		0.2%	99.3%	99.6%	96.9%		
PP & PP (EPDM)		93.6%	0.0%	0.0%	0.0%		
Filled PP		0.4%	0.0%	0.0%	0.0%		
All PP		94.0%	0.0%	0.0%	0.0%		

2001-01-0698

Masticated Rubber: A Recycling Success Story

Andrew Haber
NRI Industries Inc.

ABSTRACT

The term Masticated Rubber has become synonymous with a mixture of recycled rubber and fiber derived from passenger car tires.

Masticated Rubber has been used for over 15 years by a variety of OEMs for air & water management applications in the transportation industry. Its success is a result of being able to offer value-added solutions to recycling car tires and low price/high performance characteristics.

The objective of this paper are to provide an update on the advances made in the manufacture of Masticated Rubber compounds, to present data for typical applications and to present data showing the environmental benefits associated with using this material. Not only are fewer passenger car tires being burned for fuel or sent to landfills, but factory greenhouse gas emissions are also reduced (when compared to other rubber manufacturing processes).

INTRODUCTION

The dictionary[1] definition for "masticate" is:

"to chew or to grind (rubber, etc.) to a pulp "

In a commercial context, the term Masticated Rubber defines a family of materials that is a mixture of recycled rubber and fiber. More recently this definition has been expanded to include a family of materials that is a mixture of tire derived recycled rubber and fiber. This mixture of rubber and fiber has been engineered to align the fibers to create unique performance characteristics. Fiber adds strength, increases the modulus or stiffness, increases the overall hardness of the final product and reduces the elongation or stretching. These characteristics ensure the quality, performance and competitive advantage required to manufacture automotive parts.

NRI Industries processes over fifty million pounds of Masticated Rubber compounds annually. These materials are used to make over two hundred industrial and automotive components. Vehicle applications range from passenger cars to heavy trucks. All vehicles have needs related to Noise & Vibration and Air & Water Management and Masticated Rubber provides a cost effective solution for these requirements.

RUBBER RECYCLING

Rubber is a difficult material to recycle. Each year millions of pounds of tire plant scrap, hundreds of millions of pounds of non-tire scrap (trim, pads, belts, etc.) and over two hundred and fifty million scrap tires are discarded in North America. At NRI Industries, new technologies for recovering rubber have focused on two areas: grinding rubber to fine particle sizes and activating rubber particles to permit them to be revulcanized into the polymer matrix. These developments have resulted in several new material substitution opportunities. The rubber recycling value chain[2], which starts with scrap rubber as the input and has molded three-dimensional parts as the highest valued added output, can be described as follows:

1. Granulation: Tires are shredded and the steel, fiber and rubber are separated. The steel is sold to the scrap metal industry. The fiber is used to reinforce rubber parts in the Compounding phase. The rubber crumb is used either as a filler in the Compounding phase or processed further in the Advanced Processing phase.

2. Advanced Processing: Tire crumb is either ground to finer particle sizes or re-activated using a proprietary thermo-mechanical process[3,4].

3. Compounding: The various ingredients constituting the rubber formulation are combined and blended to give a preform that can be used for molding.

4. Molding: Material from the Compounding phase is molded into sheets, rolls and three-dimensional components.

This value chain can also be applied to the recycling of non-tire rubber parts, e.g. EPDM trim.

Rubber recycling strategies need not solely be based on using tire or rubber crumb. Materials have been developed at NRI Industries that are a mixture of tire crumb, tire fiber, revulcanized tire crumb and industrial scrap (in the form of uncured tire compound). By blending these materials together, it is possible to meet the high performance requirements, provide significant cost savings (as compared to virgin and alternate materials), exceed the recycling targets and meet the stringent quality requirements demanded by OEM manufacturers.

MASTICATED RUBBER DEVELOPMENT

Before 1990, Masticated Rubber was based on using mostly post-industrial scrap. This material, routinely referred to as Friction, is uncured tire compound that includes tire cord. Because of frequent price fluctuations related to sourcing this material, development work was initiated to find a substitute.

With the virtually unlimited supply of scrap tires and the opportunity to get tipping fees when tires are discarded, tire crumb was considered as a substitute. Processes were developed to efficiently grind tires and to separate the rubber, steel and fiber. It was found that only a small portion of the post-industrial material could be replaced, since as the concentration of tire crumb increased, the physical properties were found to decrease. Development work was initiated to try and devulcanize or re-activate the rubber crumb to allow for higher levels of substitution.

Between 1990-1994 a process was developed at NRI Industries that would re-activate the rubber crumb. This new material was called Symar™-D. Formulations were developed in which over 50% of the post-industrial material was replaced.

As these materials were being developed, it was found that a significant portion of the skeletons (cured rubber trim) from die cutting operations could be added to the formulation, thereby reducing the cost even further.

The following is an example of a typical Masticated Rubber formulation:

Formulation	
• Friction	45%
• SBR	4%
• Symar™-D	27%
• Fiber	14%
• Chemicals	10%
Recycled Content	
• Post-Consumer	40%
• Post-Industrial	50%
Recyclability	100%

As the percentage of Friction was reduced, the product quality and consistency was found to improve. This improvement was found to be due to the fact that tires are not processed in lots or batches from specific customers or types. Tire derived materials represent tires of all types and from different manufacturers. Variations, as a result, are blended out. Friction batches on the other hand, are from a specific tire plant and probably related to a specific run and specific problem causing the scrap.

The processing characteristics of Masticated Rubber differ from those of unfilled virgin rubber materials. The compound is drier and more porous. Higher processing temperatures and pressures are required as a result.

APPLICATIONS

Making products from tire scrap and end of life tires limits the color and surface appearance options available for Masticated Rubber. All products made using this technology are black and have a rough surface finish. The surface appearance could be improved by grinding the rubber crumb to finer sizes or using less fiber, but this increases the cost and reduces the benefits of adding fiber.

Masticated Rubber is an ideal material for under the hood and chassis applications. In these cases its black color blends nicely with the other chassis and engine components and the surface appearance is also not critical as the component is not visible.

The key automotive applications for masticated rubber products include:

- **Radiator Seals and Air Baffles:** These rubber products are used to fill the space between the radiator or a/c condenser and vehicle body, thereby ensuring that all air entering the front of the vehicle is directed through the radiator.

- **Splash Shields, Water Baffles and Fender Extensions:** These rubber parts are used in the fender or wheel well to reduce water spray or splash in the engine compartment. They can also be used on specific water sensitive engine or brake components as shields to reduce water or dust intrusion.

- **Close Outs and Sight Shields:** These rubber parts are often used to hide the complexity of engine compartment components behind the front of the grill of the vehicle.

- **Headlamp Hole Seals:** These rubber parts are used to seal around the rear of the headlamps against the structural sheet metal. These seals eliminate one non-functional air path for the cooling air that enters the front of the vehicle. With this path closed, air is forced to travel through the radiator and a/c condensor, resulting in improved cooling.

- **Spare Tire Protector:** These rubber parts serve as an isolator between the spare tire wheel and floor pan, thereby eliminating rubbing by the wheel on the floor pan.

- **Fuel tank pads:** These rubber parts serve as an isolator between the fuel tank and chassis, thereby reducing the transmission of noise and vibration.

Parts made for these applications must be able to withstand:

- Impact of water and road debris.
- Dynamic forces due to air flow.
- Tensile and elongational stresses caused by the jounce of the vehicle. Parts are usually fastened to different chassis or body components, which can move in different directions.
- Cold temperatures.
- Hot temperatures in the radiator section.

Rubber, especially fiber reinforced Masticated Rubber, is an ideal material for these types of applications.

The following is a list of the key automotive specifications met by Masticated Rubber products:

- GM3805M
- GM3809M
- GM6400MR
- GM6440M
- Ford WSS-M2D476-A7
- Ford WSS-M2D476-A2
- Ford WSS-M2D476-A1
- Ford ESA M2D119-A
- Chrysler MS-EE26
- Nissan NAV 607-0412

The following is data for a typical radiator seal application:

Material: F8048
Recycled Content:
- Post Consumer: 40% min
- Post Industrial: 50% min

Manufacturing Processes:
- Molded sheet or rolls for die cutting or compression molding

Physical Properties	Specification*	Result
Tensile Strength, MD	5.2 MPa min	5.4
Tensile Strength, TD	2.4 MPa min	3.0
Tear Strength, MD	26 kN/m min	34
Tear Strength, TD	52 kN/m min	60
Elongation, MD	15% min	22
Elongation, TD	40% min	50
Hardness, Shore A	82 +/-5	80
Ozone Resistance	**	Pass
Compression Set	40% max	35
Brittleness Point	-40°C	Pass
Heat Aging		
Tensile	+/-25% max	Pass
Elongation	+/-25% max	Pass

* also meets GM3805M, Ford ESA M2D119-A, Chrysler MS-EE26.
** Depends on the test specified in the customer specification.

The difference in the properties between the machine direction (MD) and transverse direction (TD) is due to preferential fibre orientation along the calendering or machine direction.

ENVIRONMENTAL IMPACT

The environmental benefits of using Masticated Rubber are substantial. They include:

- Significant reduction in the number of end-of-life passenger cars sent to landfills or burned for fuel. Over one million end-of-life passenger car tires are recycled into Masticated Rubber parts annually.
- Significant reduction in the amount of tire plant waste sent to landfills. Over twenty five million pounds of tire plant scrap is recycled annually into Masticated Rubber parts.
- Reduction in CO_2 emissions: over sixty two million kilograms less CO_2 is emitted into the atmosphere, as compared to making these parts using virgin

rubber. This estimate is based on using Rubber Association of Canada reported data for 1998.

- Plant waste recycled: All plant waste (die cutting trim, scrap, flash etc.) is recycled as grindings back into the formulation. Over fourteen million pounds of material that would otherwise be sent to landfills is recycled.
- Recyclability: All the parts made using Masticated Rubber are one hundred percent recyclable. In addition, this material does not impact the ability of a vehicle to be shredded or the quality of ASR produced.
- Similar products can be made using virgin plastic or rubber. The environmental impact of producing over sixty million pounds of virgin material can also be significant. This can be eliminated using Masticated Rubber.

COMPETITIVE PRODUCTS

Four types of materials compete with Masticated Rubber for two-dimensional applications: PVC, PE, EVA/Crumb and Crumb/Urethane. None of these materials use fiber to reinforce the material. Urethane and EVA serve as binders to hold the rubber crumb together, but neither of these alternatives is a cured product. Plastic materials also exhibit poorer cold temperature properties when compared to rubber materials.

Plastic and virgin rubber materials compete with Masticated Rubber for three-dimensional applications. Plastic materials can include PVC, PE, PP and TPO. The key disadvantage of these materials is the lack of fiber and cold temperature characteristics.

The sensitivity of Masticated Rubber product pricing to fluctuations in oil or raw material prices is significantly less than that of the other materials. With over two hundred and fifty million passenger car tires discarded annually in North America, there is virtually an unlimited supply of raw materials to draw from.

FUTURE TRENDS

Current development work has focused on expanding the range of three-dimensional parts which can be molded using Masticated Rubber. Compression molding of these complex three-dimensional shapes is possible now that the preform rheology has been tuned to allow the material to flow more uniformly in the mold.

CONCLUSION

Can mountains of discarded tires be transformed into functional, attractive, economical and environmentally sound new automobile parts? Yes, they can if they are used to make Masticated Rubber parts.

Masticated Rubber defines a family of materials that is a mixture of end-of-life tire recycled rubber, tire plant scrap and fiber, having a recycled content of at least ninety percent.

The environmental benefits of using this material include significant reduction in the amount of tires sent to landfills or burned for fuel, in the amount of tire plant scrap sent to landfills and in the amount of CO_2 emitted into the atmosphere. These benefits, along with the rubber properties and low cost make it the material of choice for Air & Water management applications.

ACKNOWLEDGMENTS

The efforts of many people at NRI Industries (previously National Rubber Company) over the past 20 years has resulted in the development of Masticated Rubber. The success of this material and the benefits from its use (financial and environmental) is due to their hard work.

REFERENCES

1. The New Lexicon Webster's Dictionary of the English Language, Lexicon Publishing Inc., NY 1987
2. A. Haber, "Engineering Automotive Products Using Recycled Rubber", SAE World Congress, Paper # 1999-01-0668, March 1-4, 1999.
3. A. Kolinski, T. Barnes, M. Schnekenburger and J. Adams, "Thermo-Mechanical Re-Activation of Tire-Rubber Crumb", Rubber Division Meeting, ACS, Nashville, Tennessee, September-October 1998, paper no. 40
4. A. Kolinski, T. Barnes, G. Paszkowski and A. Haber, "Modified Tire-Rubber Crumb as a Base Compound for Rubber Parts Manufacturing" Rubber Division Meeting, ACS, Orlando, Florida, October 1999, paper no. 106

CONTACT

Andrew Haber
NRI Industries Inc.
394 Symington Ave
Toronto, Ont, Canada M6N 2W3
(416) 652-4123
(416) 652-4214
ahaber@nriindustries.com

2001-01-0699

Improvements in Lubricating Oil Quality by an On Line Oil Recycler for a Refuse Truck Using in Service Testing

Gordon E. Andrews and Hu Li
The University of Leeds

M. H. Jones
Swansea Tribology Services Ltd.

J.Hall, A. A. Rahman and P. Mawson
Top High UK Ltd

ABSTRACT

A method of cleaning lubricating oil on line was investigated using a fine bypass particulate filter followed by an infra red heater. Two bypass filter sizes of 6 and 1 micron were investigated, both filter sizes were effective but the one micron filter had the greatest benefit. This was tested on two nominally identical EURO 1 emissions compliance refuse trucks, fitted with Perkins Phazer 210Ti 6 litre turbocharged intercooled engines and coded as RT320 and RT321. These vehicles had lubricating oil deterioration and emissions characteristics that were significantly different, in spite of their similar age and total mileage. RT321 showed an apparent heavier black smoke than RT320. Comparison was made with the oil quality and fuel and lubricating oil consumption on the same vehicles and engines with and without the on-line bypass oil recycler. Engine oils were sampled and analysed about every 400 miles. Both vehicles started the test with an oil drain and fresh lubricating oil. The two refuse trucks were tested in a different sequence, the RT320 without the recycler fitted and then fitted later and the RT321 with the recycler fitted and then removed later in the test and both without any oil change. The RT320 was also the one with the finer bypass filter. The test mileage was nearly 8,000 miles both trucks. The amount of fresh oil top up was monitored and the results corrected for this dilution effect. The results showed that the on line bypass oil recycler cleaning system reduced the rate of fall of the TBN by 23% and 49% for two trucks respectively. A 73% reduction in the rate of increase of the TAN incurred for one of the trucks. The soot in oil was reduced by ~70% on average for both trucks.

The reduction in the rate of carbon accumulation in oil was 55% for the refuse truck with heavy smoke emissions. There was a 56% reduction in iron. The rate of oxidation, nitration and sulphation of oils was significantly reduced. There was an improvement of the fuel economy of about 3%. The lubricating oil consumption was reduced by 40% for 1 micron recycler filter and 30% for 6 micron filter.

INTRODUCTION

Lubricating oil forms a significant fraction of the particulate volatile fraction and can contribute to the carbon emissions. Lubricating oil also acts as a sink for carbon emissions (1,2) and unburned diesel fuel. This can lead to the deterioration of the oil (3) and result in an increase in the particulate emissions (1). For a low particulate emissions engine, the work of Cooke (4) showed that lubricating oil might contribute more to the carbon emissions than to the solvent fraction at some engine conditions. His results showed that there was a variable influence of lubricating oil with no influence on particulate emissions for some engine conditions and up to a 250% increase with lubricating oil age for other engine conditions.

Diesel engines with low particulate emissions have a very low lubricating oil consumption. There is a concern that carbon particles may accumulate to a greater mass concentration in the lubricating oil, as there will be a reduced dilution with top up of the oil (5). High carbon in the lubricating oil may then increase the contribution of the oil to the particulate emissions through the associated higher viscosity. Andrews et al (6) have shown that a Euro 1 passenger car diesel engine accumulated carbon in the oil at a greater proportion of the carbon emissions than for an older high carbon emitting engine.

The control of combustion chamber deposits (CCD) in modern diesel engines is recognised as a part of low emission engine design and extended service requirements are making it increasingly difficult to control deposit formation (7). The primary source of piston deposits is the lubricant (8) and oxidation of the lubricant is the primary cause of deposit formation (9). In cylinder deposits consist of ash from the lubricating oil additives, carbon and absorbed unburned fuel and lubricating oil (8, 10). The CCD can be a source of wear in engines, and increased friction and hence increased fuel consumption. Crownland heavy carbon has been shown to increase oil consumption (11) and deposits have been shown to increase as piston temperature increases above 250°C (12). Deposits also increase with the soot content of the oil (12) and hence deposits can increase as the oil ages. The increased soot in oil as it ages results in an increased oil viscosity (13) and this increases the oil (14) and fuel consumption. The aim of the present work was to examine these influences for two Euro 1 refuse trucks fitted with Perkins Phazer engines and to determine the improvement in oil quality through the use of on line recycling of the oil to remove soot, wear metals, fuel and water dilution (3,15). The recycler had a fine bypass oil filter and an infrared heater to distil out water and light fuel fractions. The use of by pass filters is common in some large diesel engines, but is not usual in smaller engines of 6 litres or less.

Andrews et al (1, 6) showed that the lubricating oil age could have a significant influence on particulate emissions. Three IDI engines were tested over 100 hours to investigate the influence of lubricating oil age on the emissions. Two Ford engines, 1.6 and 1.8 litre, were low emission engines and the Petter AA1 engine was an older technology high emissions engine. For all three engines there was little influence of lubricating oil age on gaseous emissions. There was evidence in the NOx emissions for the Petter and Ford 1.8 litre engines of an action of deposit removal, which reduced the NOx and deposit, build up that increased the NOx. This was also supported by the lubricating oil additive metal analysis. The hydrocarbon emissions increased with oil age for both of the low emission engines but only the 1.6 litre Ford engine showed a similar change in the particulate VOF. The 1.8 litre engine VOF trends were dominated by lubricating oil influences, which do not contribute to gaseous hydrocarbon emissions at a 180°C sample temperature. The particulate emissions trends with oil age were quite different for the Ford 1.6 and 1.8 litre engines, with a continuous increase in emissions for the former and a decrease followed by an increase after 50 hours for the latter. The Petter engine also followed similar trends to the Ford 1.8 litre engine, although with much higher emission levels. It was shown that these trends were also reflected in the carbon fraction and unburned fuel fractions of the particulate VOF for the two Ford engines. However, the lubricating oil fraction decreased substantially over the first 50 hours for the Ford 1.8 litre engine and then remained at a stable level. The implication was that the fresh lubricating oil resulted in high unburned lubricating oil particulate VOF emissions and also generated carbon emissions. Once the volatile fraction of the lubricating oil had been burnt away in the engine, the lubricating oil VOF remained stable and the subsequent increase in the particulate emissions was due to increasing carbon emissions. The initial decrease in the fuel VOF fraction followed by an increase after 50 hours was possible due to the initial removal of CCD by the fresh lubricating oil followed by a build up of fresh deposits as the oil aged. Fuel fraction VOF can be contributed to by deposit absorption and desorption, which is a function of the extent of the CCD.

This work is concerned with a technique to keep oil clean, extend its life and reduce the increase in emission that occurs with aged oil. The above review has emphasised the importance of CCDs in emissions and lubricant quality. At the same time as diesel emissions regulations have come into force the trends in the diesel design, towards higher ring zone temperature and pressure, piston redesign for higher top rings, higher piston temperatures, and extended service interval requirements, are making it increasingly difficult to control deposit formation in the engine with traditional oil additive technology (7). Diesel deposits can be classified into two types: ring zone and piston skirt deposits (varnish). The higher ring zone temperatures (325-360°C) promote thermal degradation of the lubricating oil and unburned/oxidized fuel components producing a 'carbon' deposit. At relatively low piston skirt temperatures (200-260°C) a varnish type deposit predominates (7). The primary source of piston deposits is the lubricant and lubricant oxidation is the primary cause of deposit formation (7). Diesel engine deposits also increase with the oil consumption (8), the piston temperature (12) and the oil soot content (12). Engine deposits are a source of unburned hydrocarbons through absorption and they also act as a cylinder insulation, which increases NOx emissions because of the higher cylinder temperature (17).

Oil viscosity also increases with the oil soot content (17, 18-20), that typically accumulates at a rate of 4% of the engine particulate mass emissions (1, 2, 3, 6). This results in the oil consumption increasing with the increase in viscosity (14) and also the fuel consumption increases due to the increase in viscosity. Consequently, on line oil cleaning can result in reduced soot in the oil and in reduced fuel and water oil dilution and both of these effects can result in reduced engine deposits and a reduction in oil consumption. The reduced engine deposits can also result in reduced engine friction and hence reduced fuel consumption. Thus, improved oil quality can have additional benefits apart from the extended oil life.

THE ON LINE OIL RECYCLER AND REFUSE TRUCK TEST PROCEDURE

A method of continually cleaning the engine oil on line was investigated. This was based on the combined effects of a bypass fine oil particulate filter with a 1 or 6 micron filter element followed by an infra-red dome heater which heated the oil to 135 °C as it flowed over a conical cascade into a drain return to the oil sump. The previous work using this system (3) used a 6 micron bypass oil filter and this was the filter used in RT321 at the start of the present work. However, work was in progress to develop a finer I micron bypass filter and this had reached the prototype stage when it was decided to fit a recycler to the second refuse truck (RT320). It had originally been intended in the present work to use two nominally identical refuse trucks with the same engine and mileage. However, as will be shown in the results, the two refuse trucks had different oil deterioration rates and the accompanying emissions results showed different emissions. RT321 was operating consistently 2 A/F richer for the same duty cycle for a journey with an average A/F of 33/1. Soot accumulation in the oil and lube oil consumption for RT321 were also significantly higher. Consequently, it was concluded that RT321 was in a worse mechanical state to RT320 and the two refuse trucks could not be compared one with a recycler and one without. Thus, both refuse trucks had to be tested with and without the recycler.

RT320 was first tested without the recycler fitted and after 4,700 miles a recycler with the 1 micron bypass filter was fitted without any oil change. RT321 was first fitted with a recycler with the 6 micron bypass filter and after 5,000 miles the recycler was removed without any oil change.

5,000 miles of oil use is twice the normal oil change interval for the refuse truck. However, the oil analysis showed that the oils were still fit even for the refuse truck without the recycler fitted. This was attributed to the large amount of oil consumption and subsequent oil top up, which showed 2-5 times higher oil consumption than the FirstBus tests(15).

The aim of extended tests without the oil change for two refuse trucks was to determine the rate of deterioration of the aged oil after the recycler was removed and demonstrate in an on road test that the recycler could clean-up dirty oil and reduced its rate of deterioration and make it fit for further use. It has shown that the oil with the recycler removed needed to be changed whereas the oil with the recycler fitted was still fit for continued use. The normal oil change period on these refuse trucks was 2,500 miles, typically about every three months. The refuse trucks were operated 6 days a week with an average mileage per day of about 35 miles. These trucks had a very frequent stop-start duty cycle with many loading/unloading processes, which kept the engines in a high power output status and thus put lubricating oils in a very hostile environment.

One of the advantages of the recycler is that it provides for an improved oil quality and reduced oil consumption (3,15). The improved oil quality reduces the engine CCDs. Thus the reduced oil consumption and CCD in the combustion chamber reduce engine emissions. Authors have shown that emissions had been reduced for a Ford 1.8 litre IDI engine test as a result of improved oil quality and reduced oil consumption (16).

Other investigators of bypass filters have advocated using filters of the order of 1 micron (23-27). The initial choice in the present work of 6 micron particle size filters was based on advice from hydraulic oil filter manufacturers that oil additives could be filtered out if the filter size was too fine. Also, very fine 1 micron oil filters that were available from hydraulic oil filter manufacturers had a rather high pressure loss. If these were used in the recycler then the bypass flow would have been controlled by the filter pressure loss and the flow rate would have decreased as soot built up in the filter. A key feature of the present recycler is that the bypass oil system oil flow rate is relatively high. In the Ford 1.8 litre IDI passenger car diesel tests (3) the recycler oil flow rate was such that oil the sump volume of the oil was passed through the recycler four times an hour. The 1 micron by pass filter used in this work was developed to have a fine filtration of 1 micron particles without affecting the recycler bypass flow rate, even when loaded with soot. The present results show that this new fine filter does improve the performance of the recycler. This filter is now in production and will be used in the future commercial use and evaluation work using this system.

Bypass filters have two basic features: a high filtration efficiency and a high particulate storage capacity (25). They reduce the fine particulate matter in the oil leaving the main filter to remove the larger particulates. Stenhouer (26) showed that bypass filters removed organic material, sludge, varnish, resin, soot and unburned fuel. His work showed that over 80% of the contaminant removed by the bypass filter was organic. Bypass filters can remove the small pro-wear contaminant particles thus extending engine component life (27). The benefit of bypass filtration was the reduced wear and reduced in cylinder deposits. It is possible to arrange a combination filter whereby in one housing a main and bypass filter are arranged (25) with the flow between the two splitting according to their flow resistance.

Although a filter based bypass filter was used in the present work, centrifugal bypass filters are also quite common (27). These are of two types: powered and self powered and the latter are more common in automotive applications. A self powered centrifugal oil cleaner uses the dirty oil pressure to drive the cleaning rotor using centrifugal separation of the high density particles from the lower density oil. These contaminants collect as a hard cake on the inside of the rotor which can then either be cleaned off or disposed of as a unit (27). Centrifugal filters have an effective particle size removal below 1 micron and have a filtration efficiency that does not deteriorate with time (27). They are generally more expensive initially than filter based bypass filters. One assessment of bypass filters (28) has determined the average size rating (50% removal) of a centrifugal filter as 6-10 microns and hydraulic oil filters as 2 micron. They also estimated that it would take 30 bypass filter changes before the cost of bypass filtration exceeded that of a centrifugal filter. This could be 10 years of normal use. Hydraulic quality bypass filters were used in the present work with an average size rating of 6 and 1 microns.

EXPERIMENTAL TECHNIQUES

ENGINE SPECIFICATIONS - The two refuse trucks, coded as RT320 and RT321, had the same engine specification that is detailed in Table 1.The two refuse trucks tested operated with routine oil top ups and normal commercial duty cycles. The oil and exhaust gas samplings were taken every two weeks, at about 400 miles intervals. The same route and driver were used for each test run. However, traffic conditions were varied.

Table 1. Specifications of the engines

	Refuse truck
Type	Perkins Phazer 210Ti
Maximum Power Rating	156.5KW(210BHP) @ 2500rpm
Displacement, litre	6.0
Oil pressure, P.S.I @ 2100rpm(max)	43-49
Oil capacity, litre	18
Bore, mm	100
Stroke, mm	127
Cylinder No.	6
Compression ratio	17.5:1
Lube oil change intervals	3 months
Aspiration	Turbocharge intercooled

RT320 started the test without the recycler fitted and RT321 started the test with the recycler fitted. For RT320 after 4,700 miles after the commencement of the test with fresh oil, a recycler was fitted. Then the test continued for 3,000 miles. For RT321 after ~5,000 miles after the commencement of the test with fresh oil, the recycler was removed and the test continued for 2,600 miles without the oil being changed. As discussed above, RT320 was fitted with a recycler with a bypass particulate filter size of 1 microns and RT321 was fitted with a 6 micron filter.

FUEL AND LUBRICANTING OIL - The fuel used in the tests was commercially available standard low sulphur diesel with sulphur content ≤ 0.05%. Table 2 shows the specifications of the diesel fuel.

The lubricating oils used in tests were 15W-40 CE/SF mineral oil. The specifications for the lubricating oil are listed in Table 3.

Table 2. The specifications of diesel fuel used in the refuse truck tests

Property & unit	diesel	Test Method
Colour	≤2.5	D1500/IP196
Density, @15°C, g/ml	0.82-0.86	D4052
Flash point(PMCC), °C	≥56	ISO2719
Cetane Number	≥50	ISO5165/IP380/D613
Viscosity, @ 40 °C, cSt	2.0-4.5	ISO3104
Sulphur, %wt	≤0.05	ISO8754
Micro Carbon Residue: Residue wt on 10% Bottoms	≤0.3	ISO10370/IP398
Ash, %wt	0.01	ISP6245/IP4
Particulate Matter, mg/kg	≤24	DIN51419
Water, mg/kg	≤200	D1744
Oxidation stability, mg/100ml	≤2.5	D2274
Distillation		ISO3405
%Vol, @250 °C, rec	≤65	
%Vol, @345 °C, rec	≥95	

LUBRICATING OIL ANALYSIS - Oil was sampled periodically from the dipstick using a vacuum pump and syringe with the sample taken through flexible tubing located just below the oil surface in the sump and collected in a small glass bottle. The volume sampled was noted and taken into account when the oil consumption by the engine was determined by draining the sump at the end of the test. The oil sample was analysed for total base number (TBN) and total acid number (TAN), viscosity, soot by infra red and by thermal gravimetric analysis, wear and additive metals and additive depletion using FTIR. The oil analysis techniques used are summarised in Table 4. The oil used was SAE 15W/40 with the properties summarised in Table 3. The main oil analysis was carried out to current standard methodology at Swansea Tribology Services Ltd. and the TGA and FTIR work was carried out at Leeds University.

Table 3. Specifications of the lubricating oils

	Refuse truck test
SAE Viscosity Grade	15W/40
API Classification	CE/SF
Physical and Chemical properties	
Viscosity, @ 100 ^{o}C	14.5
mm^2/s @ 40 ^{o}C	110
Viscosity Index	
Flash point, ^{o}C	>180
Density, @ 15C, kg/m^3	<1000
TBN, mgKOH/g	13
TAN, mgKOH/g	3.7
SulphatedAsh, wt%	1.8
Elemental analysis	
Calcium, wt%	0.11
Magnesium, wt%	0.003
Zinc, wt%	0.08
Phosphorus, wt%	0.10
Manufacturer	Castrol

The lubricating oil was analysed for fuel dilution, carbon and ash content using Thermal Gravimetric Analysis (TGA). The method used a small sample size of approximately 50 mg that was placed in a small bowl hanging from a microbalance. This was then heated in an oven in a flow of nitrogen and the weight loss as a function of temperature was determined up to 600°C temperature. The nitrogen was then switched to air and the heater increased to give 610°C and any carbon in the oil was burnt away and the remaining weight was the ash in the oil.

Fuel dilution was determined by calibration of the TGA using mixtures of fuel and oil in different concentrations and different fuel boiling fractions. This was done on an oil volatility basis and the method cannot distinguish between oil degradation to give volatile low molecular weight components and fuel dilution. The calibration was undertaken with diesel/lube oil mixtures in the 0.5-5% fuel range and it was found that a temperature of 290°C for the oil and fuel used in this work enabled the fuel dilution to be determined as the weight loss up to 290°C. The boiling fractions most suitable to use were determined using pyrolysis GC of the used lube oil fractions using the fuel n-alkane distribution as the indicator of the fuel in oil distillation range. The calibration reference temperature was 290°C throughout the present work.

The carbon content of the used lubricating oil was also determined using TGA. The used lubricating oil was heated in nitrogen to 600°C where there was no further change in the volatile weight loss. Air was then added and the carbon in the oil was burnt out and the decrease in weight was the carbon fraction. Any remaining weight from the initial sample weight was the ash fraction of the lubricating oil. This technique was very similar to that used by Covitch et al (29), Ripple and Guzauskas (30) and was first used by McGeehan and Fontana (18) and is more reliable as a gravimetric measurement than the alternative optical or centrifugal methods for soot in the oil.

LUBE OIL AND FUEL CONSUMPTION

LUBE OIL TOP UP - Figures 1 and 2 have shown the accumulative oil top ups in terms of litre and percentage of the oil sump capacity for two refuse trucks. The RT320 had a higher rate of oil top ups with the recycler than without the recycler due to some oil leaking from the recycler. Without the recycler 80% of oils were added after 5,000 miles. With the recycler 100% of oils were topped over 3,000 miles test due to two substantial oil top ups as a result of a mechanical fault in the recycler which resulted oil leaking. For RT321 the oil top up rate was higher, indicating higher lube oil consumption on this truck.

The total amount of the oil added on the RT321 after about 5,000 miles of test with the recycler was 100% of the sump capacity. The quantity of the oil topped up was increased after the oil recycler was removed, showing an increased lube oil consumption without the recycler.

The mean value for lube oil top ups was calculated from the data in Figures 1 and 2 for two vehicles with and without the recycler. As the RT320 had an accidental leaking on the test with the recycler fitted, those two substantial oil top ups were excluded in the calculation. The mean value was 4.2 litre per 1,000 miles.

Table 4. Analytical methods for lubricating oils

Analytical items	Standard /method
Physical and chemical properties	
Kinematic Viscosity, @100, 40°C mm²/s	ASTM D445
TBN, mgKOH/g	D4739
TAN, mgKOH/g	D664
Soot, /cm	FTIR
Water, %	FTIR
Coolant, %	FTIR
Oxidation	FTIR
Sulphation	FTIR
Nitration	FTIR
Carbon, wt%	TGA
Ash, wt%	TGA
Fuel dilution, wt%	TGA+GC
Elemental analysis	Emission spectra
Ca Mg Zn P	
Fe Pb Cu Cd Cr Al	

LUBRICATING OIL CONSUMPTION - Lubricating oil consumption was calculated and determined by the raw top up records from the fleet. Figures 1 and 2 show a pattern of regular oil top up, except three irregular oil top ups due to mechanical faults: the first one was the last oil top up for RT321 with the recycler fitted; the other two were at 6.000 miles and 6,800 miles of oil age for RT320 with the recycler fitted. These three data points were excluded when the lubricating oil consumption was calculated.

The lubricating oil consumption has been represented on average with and without the recycler in terms of litre per kilomile (l/kmile) and g/kgfuel as listed in table 5. It has shown that the reduction in lube oil consumption by the oil recycler was 31~43% in terms of l/kmile and 30~41% in terms of g/kgfuel for two refuse trucks.

The lubricating oil consumption for RT321 was about twice as the same as that for RT320. This very high lube oil consumption resulted in high smoke as shown later. Thus this engine must have some mechanical fault, which resulted in an excessive burning of lubricating oil.

It was found that the lubricating oil consumption was much higher on the refuse truck tests than that on the FirstBus tests (2-5 times, without the recycler, 2-6 times with the recycler). This indicated that the engine technology and maintenance for the buses were better (buses were 1995 model whereas refuse trucks were 1991 model). Another factor is duty cycles. Refuse trucks were operated with a pattern of very frequent stop start cycles, which also contributed to worse oil consumption.

Table 5. Lubricating oil consumption for refuse truck tests

	RT320		RT321	
	l/kmile	g/kgfuel	l/kmile	g/kgfuel
Without the recycler	3.493	2.71	6.494	4.85
With the recycler	2.000	1.61	4.462	3.40
reduction rate %	42.7	41	31.3	29.9

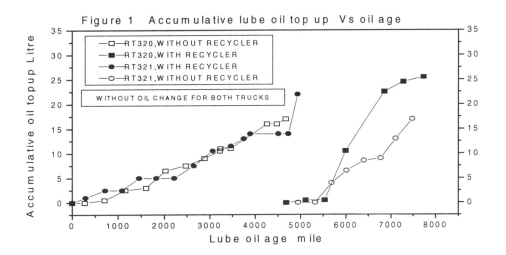

Figure 1   Accumulative lube oil top up Vs oil age

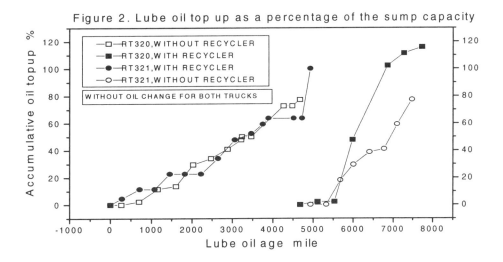

Figure 2. Lube oil top up as a percentage of the sump capacity

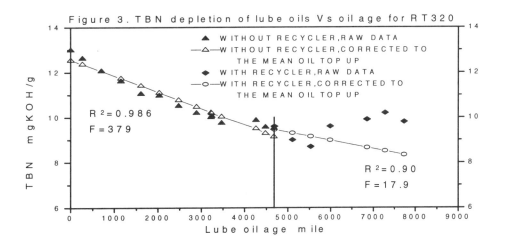

Figure 3. TBN depletion of lube oils Vs oil age for RT320

FUEL CONSUMPTION -Diesel fuel consumed was recorded by the fleet. An average value for the diesel fuel consumption was calculated from those fleet records and represented in terms of litre per kilomile (l/kmile). Table 6 shows the comparison between the results with and without the recycler for two refuse truck tests.

Table 6. Diesel fuel consumption comparison

	RT320 l/k mile	RT321 l/k mile
Without the recycler	987	1025
With the recycler	952	1004
fuel savings %	3.55	2.05

It clearly shows that the lube oil recycler has reduced the fuel consumption rate. This is attributed to the filtration of the recycler, which reduced the accumulation of the solid contaminants and formation of the deposits in cylinder. Thus the friction was reduced and energy loss due to friction was decreased.

Comparing this data with that from the FirstBus test (15), it shows that the fuel consumption for refuse trucks was 43% higher than for buses. The improvement by the recycler on fuel consumption was larger for refuse trucks. This indicated that the recycler would be more efficient if applied to older engines.

THE IMPROVEMENT IN LUBRICATING OIL QUALITY WITH THE RECYCLER AS A FUNCTION OF OIL AGE

CORRECTION FOR THE OIL TOPPING UP - The properties of lubricating oil are a function of the oil formulation, contamination and combustion condition. For the same engine running at a similar driving condition, the deterioration of the engine oil is a function of oil age and amount of the oil topped up. The ageing of the oil is accompanied by the gradual contamination of the oil by blow-by gases, which bring soot, water and unburned fuel into the oil. Direct contamination with soot and fuel occurs in the lubricating oil film on the liner inside the cylinder. The oil top ups have a direct dilution effect on the oil quality parameter such as soot content and fuel dilution. Therefore all the measurements of the oil's properties in the refuse truck tests were analysed as a function of the oil age and accumulative amount of the oil topped up, using the computational multiple regression tool. Then the change of the oil properties can be compared under the same oil top up rate as a function of oil age. The main correction was for the very large oil top ups that were

required by component failures in the lubricating oil system, as discussed above. The procedure essentially derived a uniform rate of oil deterioration for the different oil quality parameters. This then allowed the change in the rate of oil quality deterioration to be determined.

The regression equations for oil properties are in the form of following:

oil property = a + b*G+c*Vtp

Where:

a----the intercept value

b,c ----the slope value

G----lube oil age, mile

Vtp----accumulative amount of the oil topped up, litre.

This is a two variables linear regression formula. The results from this equation have a very good correlation of the experimental data. To compare the rate of the oil deteriorate with the oil age, the rate of oil top up has to be fixed so as to see the oil's decay under an equal oil top up rate. An average rate of the oil top up from the two refuse truck tests was 4.2 l/kmile. This was used to produce oil top up corrected data as a function of the oil age in all following results. The linear fit to the experimental data that resulted from this procedure is marked on all the graphs of the oil quality as a function of oil age. This enables the linear regression line to be compared with the raw data.

The effect of the oil recycler on oil quality from this procedure is the comparison of the deterioration in the rate of change of an oil parameter with oil age. The ratio of the results with and without the oil recycler then gives a percentage improvement in that parameter due to the oil recycler. These results are summarised in Table 7 for all the oil parameters that were determined. Each parameter is discussed in detail below. However, it can be seen that for every oil parameter measured there was a very significant reduction in the rate of deterioration with oil age when the recycler was used.

TBN AND TAN - The depletion of basic component in lubricating oil additives during the use was measured by the determination of TBN (Total Base Number). Figures 3 and 4 show depletion rate of oil TBN for two refuse trucks respectively.

For RT320, the value of TBN decreased from 12.4 mgKOH/g to 9.2 mgKOH/g after 4,700 miles without the recycler. The rate of depletion was 0.7 mgKOH/g per kilomile. With the recycler fitted, TBN value decreased 1.1 mgKOH/g during a further 3,100 miles of use of the oil. The rate of depletion in TBN was reduced to 0.4 mgKOH/g per kilomile after fitting the recycler to aged oil.

Table 7. Summary of the comparison on oil qualities with and without the recycler on two refuse truck tests

Parameters	RT320		reduction by recycler %	RT321		reduction by recycler %
	w/o re	w. re		w/o re	w. re	
TBN depletion rate, % /kmile	5.7	2.9	49	5.2	4	23
TAN increase rate, %/kmile	3	-3.9		16	4.4	73
Soot increase (IR), abs/kmile	3.6	0.65	82	21	7.4	65
Carbon accumulation, %/kmile	0.19	-0.16		0.56	0.25	55
Iron increase, ppm/kmile	5.3	-2		22.8	9.6	56
Oxidation, abs/kmile	2.1	-0.32		3.2	2	37.5
Nitration, abs/kmile	0.96	0.06	94	4	2.2	45
Sulphation, abs/kmile	2	0.16	92	5	4	25

N.B. w/o re--without the recycler.   w.re--with the recycler

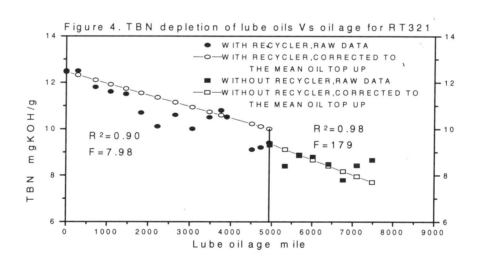

Figure 4. TBN depletion of lube oils Vs oil age for RT321

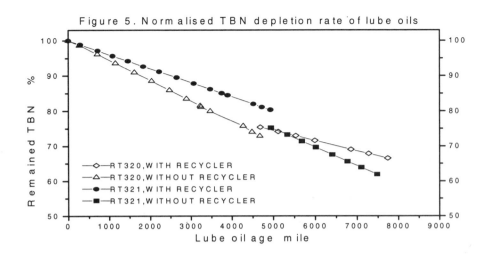

Figure 5. Normalised TBN depletion rate of lube oils

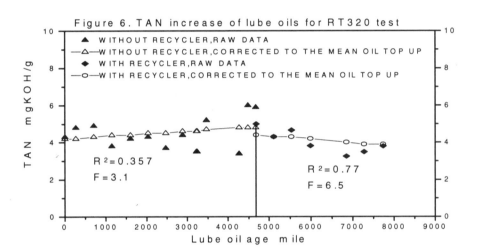

Figure 6. TAN increase of lube oils for RT320 test

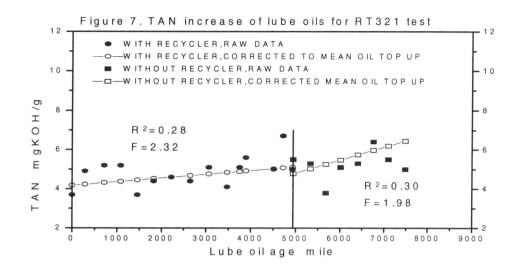

Figure 7. TAN increase of lube oils for RT321 test

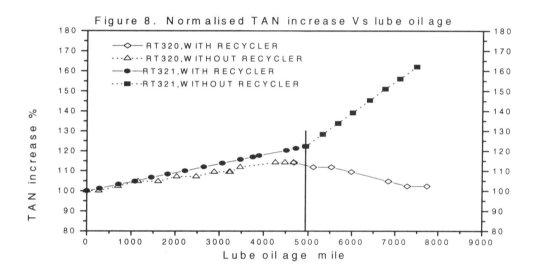

Figure 8. Normalised TAN increase Vs lube oil age

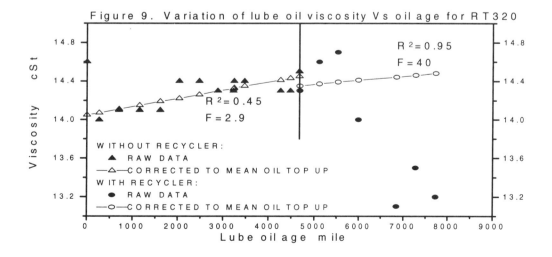

Figure 9. Variation of lube oil viscosity Vs oil age for RT320

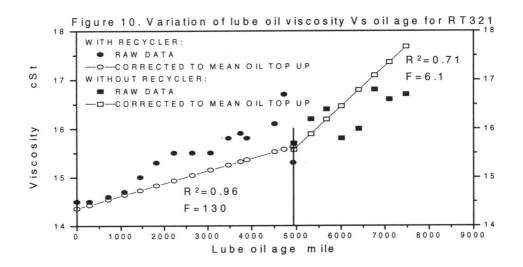

Figure 10. Variation of lube oil viscosity Vs oil age for RT321

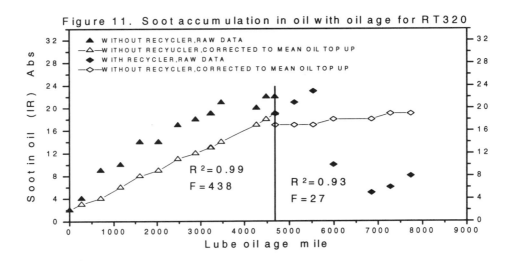

Figure 11. Soot accumulation in oil with oil age for RT320

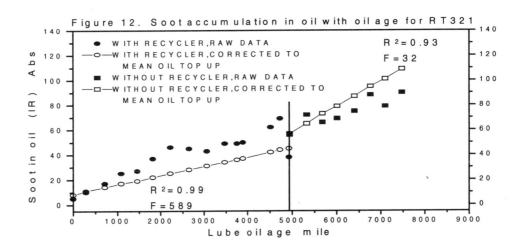

Figure 12. Soot accumulation in oil with oil age for RT321

132

An improvement of 43% in the rate of depletion of TBN can been seen with the recycler fitted.

For RT321, with the recycler the value of TBN decreased by 2.5 mgKOH/g from fresh oil during 5,000 miles of use of the oil. The rate of decrease was 0.50 mgKOH/g per kilomile. After the recycler had been taken off, the TBN value of the oil was reduced by 1.7 mgKOH/g in a further 2,500 miles of use. The rate of decrease was 0.68 mgKOH/g per kilomile. An increase of 36% in the rate of depletion of TBN with the recycler off or a reduction of 27 % on the TBN depletion rate with the recycler fitted was seen.

The TBN depletion data was normalised to the initial values and displayed in Figure 5. All the data was with mean oil top ups. it can be seen that for RT320 the TBN of the oil remained 73% of its original value after 4,700 miles of use without the recycler whereas only 9% further decrease after the recycler being fitted after another 3,100 miles. The rate was 5.7% without the recycler and 2.9% with the recycler per kilomile, a 49% improvement by the recycler. For RT321 the TBN remained 80% of its original value after 5,000 miles of use of the oil with the recycler. After the recycler was taken off, the remnant TBN was 67% at the end of a further 2,500 miles of use of the oil. The rate was 4% with the recycler and 5.2% without the recycler per kilomile for RT321. Hence the improvement on the depletion rate of TBN was 23%.

The TAN data fluctuated with oil age and correlation coefficients were not very satisfactory. This was due to the difficulties in the measurement of used oil TAN. Nevertheless, the trend could be found to make comparisons. Figures 6 and 7 show the TAN increase with oil age for two refuse trucks. For RT320, without the recycler the TAN values were increased slightly (0.8 mgKOH/g). In comparison, the TAN was decreased slightly with the recycler fitted due to that the declining of TAN from the addition of the fresh oil surpassed the increase from the accumulation of acidic materials. This meant that the TAN values would not have any increase if the oil top up was kept at a rate of 4.2 litre per kilomile (mean top up rate). For RT321, the TAN values increased by 0.9 mgKOH/g from fresh oil after 5,000 miles with the recycler. After the recycler was taken off, an increase of 1.7 mgKOH/g in a further use of 2,500 miles was observed. The rate of increase was 0.18 mgKOH/g with the recycler and 0.68 mgKOH/g without the recycler per kilomile.

Figure 8 shows the normalised TAN increase. It can be clearly seen that the increase rate of TAN was significantly larger without the recycler, comparing with the recycler fitted on both vehicles. The increase rate was 3% without the recycler and -3.9% with the recycler per kilomile for RT320, and 4.4% with the recycler and 16% without the recycler per kilomile for RT321. The reduction in the rate of TAN increase by the recycler was 73% for RT321.

VISCOSITY - The viscosity of oils is presented in Figures 9 and 10 for RT320 and RT321 respectively. For RT320 without the recycler there was a minor increase in the oil viscosity (0.4 cSt after 4700 miles). With the recycler fitted, there was a deep decrease in viscosity after 5,500 miles of oil age, which was due to the large amount of oil topped up with a lower viscosity. The regression analysis showed that under a mean oil top up rate the oil viscosity will be increased by 0.13 cSt in 3,000 miles of use of the oil. In general, the oil viscosity did not increase significantly for RT320 through the whole test.

For RT321 with the recycler, the viscosity of the oils increased from 14.4 to 15.6 cSt after 5,000 miles of use. With recycler removed from this truck the viscosity of the oil reached 17.7 cSt after a further 2,500 miles of use. This value, 17.7 cSt, had exceeded the maximum limit of the viscosity grade (SAE J300), which should be less than 16.3 cSt (1). This indicated that an oil change should be considered in actual use as some oil manufacturers recommend that one of the criteria for changing the oil is when the oil viscosity at 100 ^{o}C exceeded the SAE grade limit (3). The increase rate in terms of cSt per kilomile was 0.24 with the recycler and 0.84 without the recycler. The reduction by the recycler is a factor of 3.5.

SOOT/CARBON CONTAMINATION IN OIL - The soot or carbon contamination in the diesel engine oils is one of the major causes for oil deterioration, especially in old engines. The refuse trucks carry out a very frequent stop start working cycles daily and this makes the lubricating oil work under severe conditions.

The carbon contamination in oil was measured by two methods: FTIR and TGA. The FTIR measures the soot content in oil by measuring the spectra absorbency of the soot in oil. TGA determines the total carbon content in oil for both fine soot and coarse particles.

The soot in oil measured by FTIR was shown in Figures 11 and 12. For RT320 without the recycler, the soot in the oil increased from the 2 Abs of fresh oil to 19 Abs after 4,700 miles. The initial value for fresh oil was not zero because the fresh oil was inevitably mixed with the remnant used oil in the engine. After the recycler had been fitted to RT320, there was almost no increase in soot. For RT321, with the recycler the increase in soot during 5,000 miles of use was 37 Abs. The higher initial soot value (8 Abs) for 0 hour oil sample indicated more residue of used oils in the engine. After the recycler had been taken off, the soot increased by 52 Abs in 2,500 miles. Thus the rate of increase in soot in terms of Abs per kilomile was calculated as follows:

	RT320	RT321
without the recycler	3.6	21
with the recycler	0.65	7.4
reduction by the recycler	82%	65%

The data indicated 65% and 82% improvements on the reduction of soot accumulation rate by the recycler. The greater improvement for RT320 was achieved due to that the finer filter media (1 micron pore size) was used.

The total carbonaceous materials in oil were determined by TGA in terms of mass percentage, as shown in Figure 13 and 14. For RT320, without the recycler the carbon was accumulated to 0.9 %wt after 4,700 miles and showed a declined trend after the fitting of the recycler. This meant that the rate of accumulation of carbon in the oil was lower than the reduction rate due to the dilution of oil under an assumed mean oil top up rate. As a result, it revealed a declined trend with the oil ageing.

The accumulation of carbon in the oil for RT321 also showed a reduction by the recycler. The carbon in the oil was accumulated in a rate of 0.25%wt with the recycler and 0.56%wt without the recycler per kilomile of oil age.

Both soot and total carbon measurements revealed that the engine oil on the refuse truck RT321 had a significantly higher carbon contamination. This indicated that the combustion process in this engine was less complete. The air/fuel ratio measurement confirmed that the RT321 had a richer combustion condition (reported in a separate SAE paper about emissions for this refuse truck trial).

ASH CONTENT IN OIL - The ash content in oil represents the metals from lubricant additives and engine wear. It is primarily related to the lubricant additive package. As the diesel engine oils require a high detergency at high temperature a considerable amount of metal detergent is added. The ash value is influenced by the detergent content in oil to a large extent.

The ash content can also reflect the engine wear as it is also related to the wear metals in oil.

The ash in oil will be left when the lubricating oil is burnt in the combustion chamber, which can cause pre-ignition. The ash can contribute to crownland deposits above the piston rings and may lead to valve leaking so as to cause seat burning. The ash in oil also contributes a remarkable proportion to particulate emissions.

The TGA technique was used to determine the ash content in oil. The oil was firstly heated in the atmosphere of nitrogen up to 600°C and then air was introduced to burn the carbonaceous materials. The mass left after this was the mass of ash in the oil, which are mostly metal oxides. Figures 15 and 16 show the ash in oil with oil age for two refuse truck tests.

For RT320 without the recycler, the ash in the oil increased in the first 1,000 miles and then stabilised at around 2.0%wt. After the recycler had been fitted, the ash in the oil decreased. It declined to the ash level of the fresh oil after 2,000 miles of use of the recycler. This indicated a reduction in wear metals and lubricating oil consumption.

For RT321 with the recycler, the ash content in the oil was kept at around 2%. After taking off the recycler, the ash in the oil was slight increased. The ash in the oil would increase more significantly if the mean oil top rate was applied without the recycler. The increase in ash accumulation after the recycler being taken off implied an increase in the wear metals in the oil and higher lubricating oil consumption.

WEAR METALS IN OIL - Wear metals in oil were measured by emission spectra. The iron is a primary wear metal in engine oil and the level of iron content is a general indicator for wear in an engine. The main source for iron is from liner scuffing and general wear.

For RT320 test the iron in the oil increased by 25 ppm after 4,700 miles of oil age without the recycler, as shown in Figure 17. The increase rate was 5.3 ppm per kilomile. After the fitting of the oil recycler, the iron in the oil decreased by 2 ppm per kilomile even under a mean oil top up rate. This decrease in iron meant that if the mean oil top up rate was applied, the accumulation of iron in the oil would be slower than reduction due to the dilution of the oil top up. Therefore, the difference in accumulation of iron with and without the recycler is quite apparent.

For RT321 test the iron in the oil increased by 9.6 ppm per kilomile with the recycler and by 22.8 ppm per kilomile without the recycler, as shown in Fig.18. An improvement of 58% in the reduction of iron accumulation using the recycler has been achieved.

The lead and copper contents in oil are known to be the bearing metal indicators. Both refuse trucks did not show any indication of serious bearing wear. For RT320 without the recycler, the lead and copper were slightly increased to 6 ppm and 4 ppm respectively after 4,700 miles of use of the oil. With the recycler fitted there was an increase in lead after 7,000 miles of oil age. The level of lead, however, was still well below the typical value 20~40 ppm (1). The copper level had a radical increase at the same point, which was due to a coolant leak.

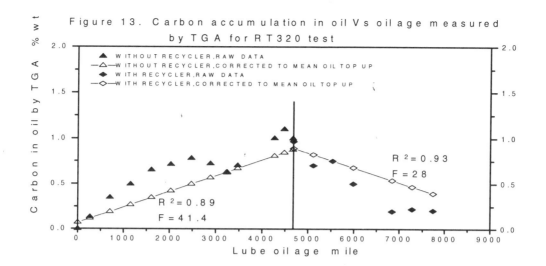

Figure 13. Carbon accumulation in oil Vs oil age measured by TGA for RT320 test

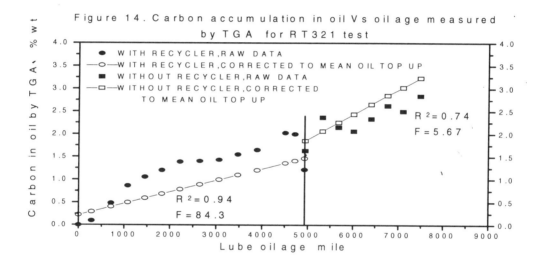

Figure 14. Carbon accumulation in oil Vs oil age measured by TGA for RT321 test

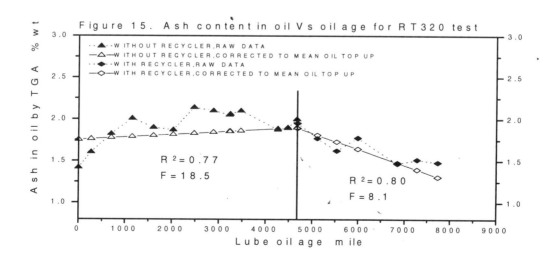

Figure 15. Ash content in oil Vs oil age for RT320 test

135

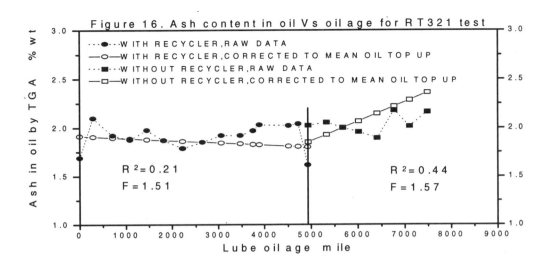

Figure 16. Ash content in oil Vs oil age for RT321 test

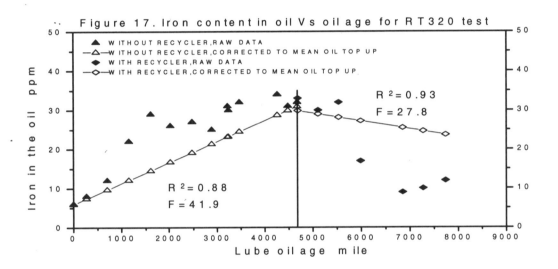

Figure 17. Iron content in oil Vs oil age for RT320 test

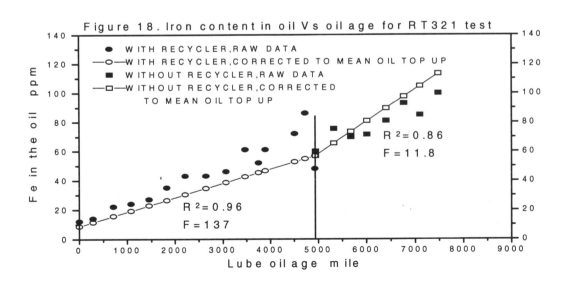

Figure 18. Iron content in oil Vs oil age for RT321 test

For RT321 test the increase rate of lead in the oil was 0.7 ppm/kilomile with the recycler and 4.8 ppm/kilomile without the recycler. The rate of increase in copper was at a similar level of 1.9 ppm/kilomile both with and without the recycler in the case of mean oil top rate.

It should be pointed out that the level of lead and copper during the whole test period for two refuse trucks were well below the warning limits, which are 50 ppm for lead or copper. They were also below the typical value: 15-25 ppm for copper and 20-40 ppm for lead. This indicated that the bearings friction was well controlled.

ADDITIVE DEPLETION:

ELEMENTAL ANALYSIS - There was no good correlation between Ca, Zn and P with lubricating oil top ups and oil age. As a result, the raw data was used to show the variation of additive elements as a function of oil age as shown in Figures 19 and 20.

For RT320 test, there were two peaks in calcium and zinc contents before the recycler was fitted and one small peak after the recycler was fitted. The content of phosphorus was stable both with and without the recycler. For RT321 test, there was a large peak in calcium and zinc after the oil recycler had been taken off, although some small peaks appeared before that. These significant peaks in additive metals indicated possible deposits burning off or dissolving in the oil and thus the calcium and zinc stored in the sediment were released into the oil. As the phosphorus is a non-metal element and not stored in the deposits and therefore its content did not change along with the calcium and zinc.

ZDDP DEPLETION - The rate of depletion of ZDDP in oil was measured by FTIR and shown in Figures 21 and 22 for two refuse trucks respectively. For RT320, the rate of depletion of ZDDP was 13% per 1.000 miles without the recycler and -1% with the recycler fitted, which showed that the content of ZDDP in oil with an average oil top up was actually increasing and indicated the decrease of ZDDP with oil age was slower than the increase due to oil top up.

For RT321, the rate of depletion of ZDDP was 5% per 1000 miles with the recycler fitted and 16% per 1000 miles without the recycler. A reduction of 69% in ZDDP depletion was achieved by the recycler.

The refuse truck RT321 showed a higher rate of depletion of ZDDP than RT320, which was associated with high smoke emissions.

OXIDATION, NITRATION AND SULPHATION OF THE OILS:

OXIDATION OF THE OILS - The engine oils are oxidised in use due to the high temperature and severe working environment. The duty cycles have a direct influence on oil oxidation. The refuse trucks carry out a very frequent stop start cycle and long time in high power output due to loading/unloading the rubbish. This would lead to a higher rate of oxidation of the oil, compared to the test results from FirstBus

Figure 23 shows the oxidation of the oil with oil age for RT320 test. The 10 units were increased after 4,700 miles of use without the recycler whereas one unit of decrease occurred with the recycler fitted under a mean oil top up rate. The rate of increase in oil oxidation was 2.1 Abs/kilomile without the recycler and -0.32 Abs/kilomile with the recycler.

The oxidation of the oil for RT321 test has been shown in Fig.24. The increase with the recycler was 10 Abs after 5,000 miles with the recycler and 8 Abs after 2,500 miles without the recycler. The rate of increase was 2 Abs/kilomile with the recycler and 3.2 Abs/kilmile without the recycler. Thus the improvement on the reduction of oil oxidation rate by the recycler was 37.5%.

The RT321 had shown a higher oxidation rate than RT320. This indicated that the engine oil in the RT321 underwent a more severe environment.

NITRATION OF THE OILS - The lubricating oil in the diesel engines is nitrified due to the contamination of nitrogen oxides from the combustion chamber. The level of nitration of the oil can indicate the tendency of deposits or varnish formation.

Figure 25 shows the nitration of the oil for RT320 test. It was increased by 4.5 Abs in 4,700 miles of use without the recycler and 0.2 Abs in 3,100 miles of use with the recycler in the case of a mean oil top up rate. The rate of increase in nitration was 0.96 Abs per kilomile without the recycler and 0.06 Abs per kilomile with the recycler. The reduction in the rate of nitration by the use of the recycler is 94%.

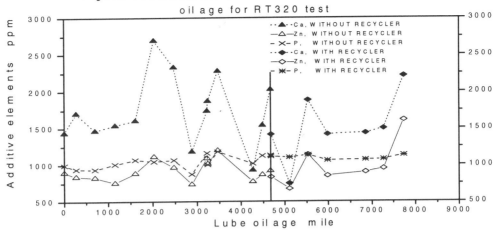

Figure 19. Contents of additive elements in oil Vs oil age for RT320 test

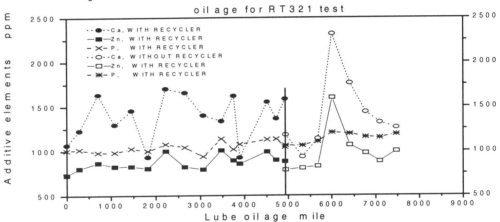

Figure 20. Contents of additive elements in oil Vs oil age for RT321 test

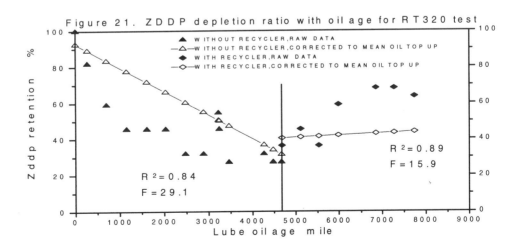

Figure 21. ZDDP depletion ratio with oil age for RT320 test

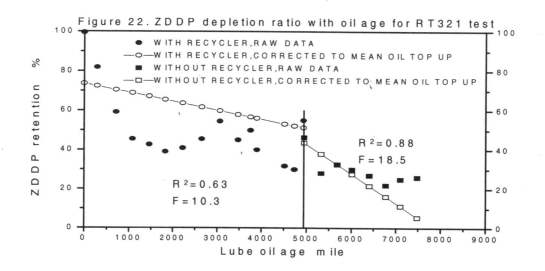

Figure 22. ZDDP depletion ratio with oil age for RT321 test

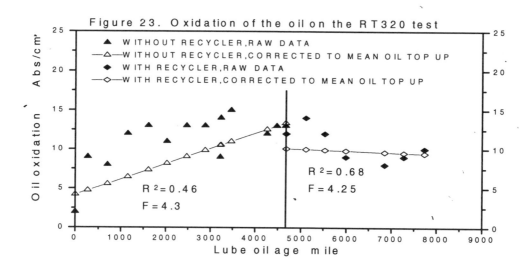

Figure 23. Oxidation of the oil on the RT320 test

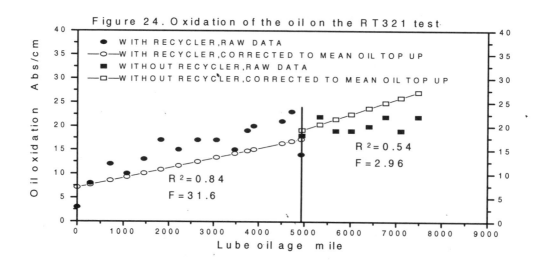

Figure 24. Oxidation of the oil on the RT321 test

Figure 26 shows the RT321 test results. The nitration of the oil showed an increase of 11 Abs with the recycler in 5,000 miles of use and 10 Abs without the recycler in a further 2,500 miles of oil age. Thus the rate of increase in oil nitration was 2.2 Abs per kilomile with the recycler and 4 Abs per kilomile without the recycler. An improvement of 45% in the reduction of oil nitration was achieved by the use of the recycler.

SULPHATION OF THE OILS -The refuse trucks use modern low sulphur diesel with ≤ 0.05% sulphur in the whole test. So there was no influence of fuel change on sulphation of the oil.

Figure 27 shows the sulphation of the oil for RT320 test. The increase of sulphation was 9.5 Abs without the recycler and 0.5 Abs with the recycler. The rate of increase was 2 Abs/kilomile without the recycler and 0.16 Abs with the recycler. Thus the reduction of sulphation by the oil recycler was 92%.

For RT321 test, the sulphation of the oil increased by 20 Abs in 5,000 miles with the recycler and 12.5 Abs in 2,500 miles without the recycler, as shown in Fig.28. The rate of increase was 4 Abs/kilomile with the recycler and 5 Abs/kilomile without the recycler. The reduction in oil sulphation rate by the oil recycler was 25%.

It has been shown above that the reductions on the oil oxidation, nitration and sulphation by the oil recycler were more significant for RT320 than that for RT321. This was considered to be due to the finer filter in the recycler for RT320. So it indicated that the finer filter has a better effect on maintaining the oil quality.

WATER IN OIL AND FUEL DILUTION - Figures 29 and 30 show the water content in oil measured by the infrared technique. For RT320 without the recycler, the water content in the oil fluctuated significantly. The maximum value for water content in the oil reached nearly 0.2%, which was reckoned as a critical value for engine oils. The water in the oil was 0.16%(1600 ppm) at the end of 4,700 miles test without the recycler. After the recycler was fitted, the water in oil was reduced from 0.16% to 0.105% instantly and continued to decrease to 0.06% after 800 miles of run with the recycler fitted. The water in the oil was then increased gradually with oil age. For RT321, with the recycler fitted the water in the oil decreased with oil age from 0.14% of fresh oil to 0.07% at the end of 5,000 miles of test with some fluctuations. After the recycler was taken off, the water in the oil continued to fall until 6,500 miles of oil age, followed by an increase. This fall in water content in oil was difficult to explain. Nevertheless, the RT320 did show an effect on removing the water from the oil with the fitting of the

oil recycler and verified that the water in oil could be accumulated to a substantial amount with no recycler fitted.

The fuel dilution can be determined by TGA technique, which was shown in reference (3) on the Ford 1.8L IDI engine test. However, the determination of the fuel dilution in oil on the refuse truck tests was interfered by the oil top ups, similar to the case of FirstBus tests (15). The distillation range of the oil changed because different batches of the oils were used to top up. The weight loss of the oil samples at a calibrated temperature varied a lot and surpassed the range of calibration. Therefore the fuel dilution could not be determined by the TGA technique.

The infrared technique was also used to determine the fuel dilution of the oils. There is a limit of 2% for infrared technique. Therefore infrared technique can only detect the fuel dilution above 2% quantitatively. The data for the whole test period on the refuse truck tests revealed the fuel dilution was below 2%.

The Gas chromatography was used to analyse oil samples qualitatively, as shown in Figures 31-35. The fresh and used oil samples with/without the recycler were analysed. There was a slight increase in a range of C16-C20 peaks for used oil samples both with and without the recycler, compared to the trace of the fresh oil. It indicated that no notable fuel dilution occurred in the tests, even without the recycler.

FAST RESPONSE OF THE RECYCLER

The oil samples were taken just before and 8 miles after the fitting/taking off of the oil recyclers for two refuse trucks. The analysis results showed there was a fast response of the recycler on oil quality with and without the recycler. Table 8 summarised the results.

The recycler was fitted to the engine after 4,700 miles of use of the oil without the recycler for RT320. All parameters, except TBN, of the oil quality showed an immediate drop by the recycler. This improvement was achieved just 8 miles after the fitting of the recycler. Similarly for RT321, all aspects of the oil quality had been getting worse just 8 miles after taking the recycler off. The differences between with and without the recycler for RT321 were larger than that for RT320. This was because the engine oil in RT321 deteriorated faster than the oil in RT320. The possible source for this was the combustion process in RT321 was more incomplete and thus more contamination of the oil.

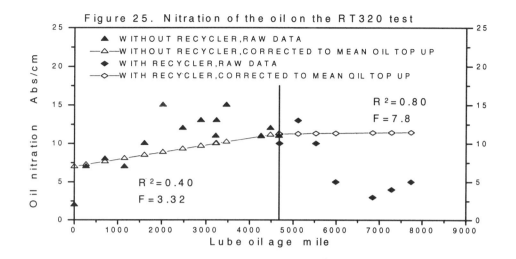

Figure 25. Nitration of the oil on the RT320 test

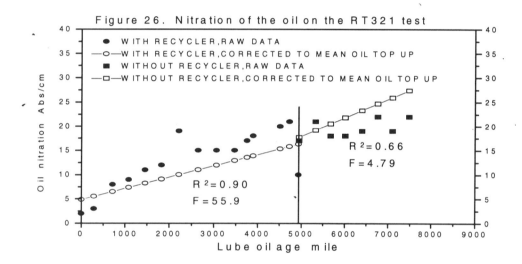

Figure 26. Nitration of the oil on the RT321 test

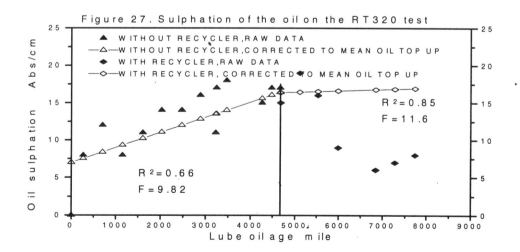

Figure 27. Sulphation of the oil on the RT320 test

141

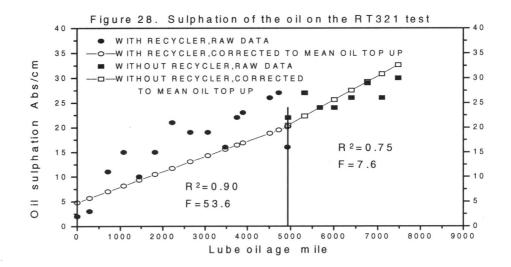

Figure 28. Sulphation of the oil on the RT321 test

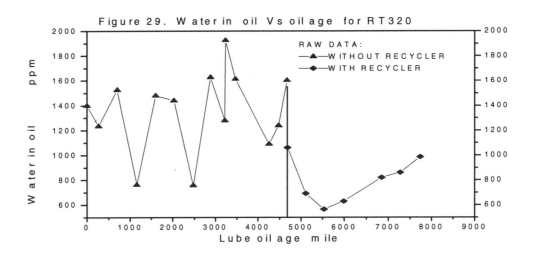

Figure 29. Water in oil Vs oil age for RT320

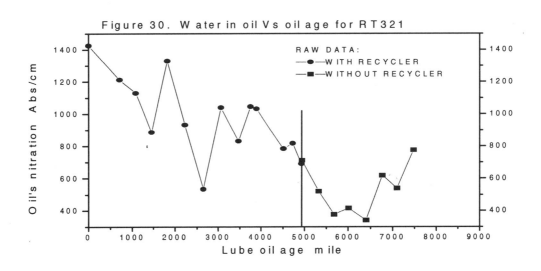

Figure 30. Water in oil Vs oil age for RT321

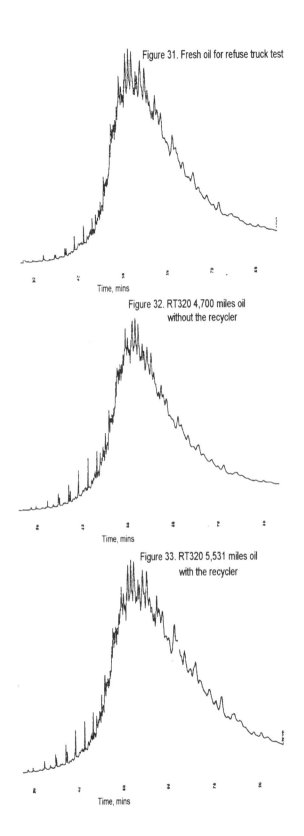

Figure 31. Fresh oil for refuse truck test

Figure 32. RT320 4,700 miles oil without the recycler

Figure 33. RT320 5,531 miles oil with the recycler

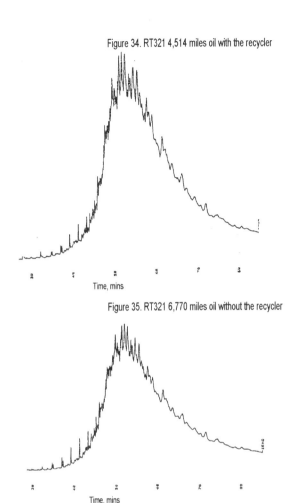

Figure 34. RT321 4,514 miles oil with the recycler

Figure 35. RT321 6,770 miles oil without the recycler

The immediate function of the oil recycler indicated the importance of filtration on improving oil quality. The flow rate of the oil through the recycler was about 0.72 l/min. Thus the recycler recycled all the oil in the sump in about 25 minutes. As the emission sampling process took about half an hour, all the oil in the sump had already been recycled once for RT320 before the first oil sample was taken after the recycler had been fitted.

ECOMONIC AND ENVIRONMENTAL BENEFITS

Refuse truck tests showed that the lubricating oil quality was well maintained with time with the recycler fitted. The better oil quality extended the oil life. The Table 7 showed that the rate of oil deterioration for most parameters were reduced by at least 50% with the recycler fitted, which indicated a doubled oil life at least. Moreover, the recycler reduced oil consumption by 30~41% (Table 5), which resulted in a further oil saving.

The improved lubricating oil quality also resulted in less deposits in the cylinders and reduced fuel consumption by 2.05 and 3.55% (Table 6) for two refuse trucks. Table 9 estimated the economies generated by the recycler for refuse trucks on average. There are four oil changes for a year operation normally. With the recycler fitted, the oil life is doubled at least and therefore the maximum oil changes are twice a year. The recycler not only extended the oil life but also reduced the amount of lube oil topped up (Table 5 showed a reduction of 37% on average). The reduction on the fuel consumption is 2.8% on average by the recycler (Table 6). The fuel savings are depending on the travel distance of the vehicles. It is assumed that a refuse truck travels 10,000 a year. So the total savings on lubricating oil and fuel are $245.6 per year for a refuse truck. Besides lube oil and fuel savings, there are other savings such as reduced maintenance cost and extended engine life. Therefore the payback time for investment is estimated less than one year for refuse trucks.

The extended oil life and reduced lube oil consumption reduced the amount of waste oils by at least 50%. Improved oil quality reduced exhaust emissions, especially black smoke and particulate emissions (reported in a separate SAE paper). These are obviously the environmental benefits generated by the oil recycler.

Table 8. The change of oil properties just before and after the fitting/taking off of the recycler

Properties	RT320		RT321	
	without recycler	after recycler fitted	with recycler	after recycler taken off
TBN mgKOH/g	9.6	9.6	9.4	9.3
TAN mgKOH/g	5.9	5.8	5.0	5.5
V100 cSt	14.5	14.3	15.3	15.7
SOOT(IR) Abs	22	19	38	56
Carbon %wt	1.0	0.96	1.2	1.6
Ash %wt	2.00	1.95	1.6	2.0
Fe ppm	33	32	48	60
Oxidation Abs	13	12	14	18
Nitration Abs	11	10	11	17
Sulphation Abs	17	15	16	22

CONCLUSIONS

Two refuse trucks with a similar age were selected and the tests were carried out to evaluate the influence of the TOP-HIGH lubricating oil recycler on emissions and lube oil quality as well as lubricating oil and fuel consumption. The two refuse trucks were 1991 model fitted with Perkins Phazer 210Ti series turbocharger intercooled engines with EURO-I emission compliance. These two refuse trucks carried out similar duties and used the same mineral lubricating oils and low sulphur diesel fuels. The lubricating oil and emissions samples were taken every two weeks or 400 miles on average and analysed. The results are showing that with the recycler:

The lubricating oil consumption has been reduced by 30~41% in terms of g/kgfuel.

The diesel fuel consumption has been reduced by 2~3.5% in terms of L/mile.

A reduction of 27~43% on TBN depletion rate has been achieved.

The increase of oil TAN has been reduced significantly.

The rate of increase in soot and total carbon in oil has been reduced dramatically. The rate of soot accumulation in oil was 65-82% lower with the oil recycler, compared to the results without the recycler. The carbonaceous materials in oil decreased with oil ageing after fitting the recycler for RT320.

The rate of accumulation of wear metals in oil has been reduced by 58% at least.

The oxidation, nitration and sulphation of the oils have been slowed down significantly.

ONGOING AND FUTURE WORK

The further development for more efficient heating dome and greater capacity filter media is ongoing. The tests on tractors and petrol vehicles are in progress. The trials for railway and marine diesel engines are planned.

ACKNOWLEDGEMENTS

We would like to thank Top High UK for a research scholarship to Hu Li and for a research contract. We would like to thank Manchester City Council, Gorton depot for their co-operation in this work and in particular to John Stanway, General Manager and Joe Stephens, Works Manager. The technical assistance was given by Ian Langstaff and Geoff Cole.

Table 9. Economy analysis for refuse trucks (for one year operation)

	without recycler	with recycler	savings
Lube oil change l	4 times/year 4x18= 72	2 times/year 2x18=36	36Lx£0.35/L = £12.6
Lube oil top up l	42[2]	26.5	15.5Lx£0.35 /L=£5.4
Lube oil Total l	114	63	£18
Fuel	1016[3]L/kmile x10=10160 L	10160x97.2% =9875.5 L	284.5x£0.8 /L=£227.6
Total saving for lube oil and fuel			£245.6
Other savings	Less engine wear and thus less equipment and maintenance cost. Fewer main oil filters required. Reduced disposal costs for waste oil and filters.		
Outlay	Initial cost for oil recycler installation and thereafter replacement of fine filters.		

N.B. 1. Assume 10,000 miles of travel a year.
2. Average rate for two refuse trucks.
3. Average fuel consumption for two refuse trucks.

REFERENCES

1. Andrews, G.E., Abdelhalim, S. and Williams, P.T., The influence of lubricating oil age on emissions from an IDI diesel, SAE Paper 931003, 1993.
2. Sun, R., Kittleson, D.B. and Blackshear, P.L., Size distribution of diesel soot in the lubricating oil, SAE Paper 912344, 1991.
3. Andrews, G.E., Abdelhalim, S. and Williams, P.T., The influence of lubricating oil age on emissions from an IDI diesel, SAE Paper 931003, 1993.
4. Sun, R., Kittleson, D.B. and Blackshear, P.L., Size distribution of diesel soot in the lubricating oil, SAE Paper 912344, 1991.
5. Andrews, G.E., Li Hu, Xu, J., Jones, M., Saykali, S., Abdul Rahman, A. and Hall, J., Oil quality in diesel engines with on line oil cleaning using a heated lubricating oil recycler, SAE Paper 1999-01-1139, 1999
6. Cooke, V.B., Lubrication of low emission diesel engines, SAE Paper 900814, 1990.
7. Cartellier, W. And Tritthart, P., Particulate analysis of light duty diesel engines with particular reference to the lube oil particulate emissions, SAE Paper 840418, 1984.
8. Andrews, G.E., Abdelhalim, S.M. and Hu Li, The influence of lubricating oil age on oil quality and emissions from IDI passenger car diesels, SAE Paper 1999-01-1135,1999.
9. Hutchings, M., Chasan, D., Burke, R., Odorisio, P., Rovani, M. and Wang, W., Heavy duty diesel deposit control – prevention as a cure, SAE 972954, 1997.
10. Graham, J.P. and Evans, B., Effects of intake valve deposits on driveability, SAE Paper 922200, 1992.
11. Andrews, G.E., Abdelhalim, S., Abbass, M.K., Asadi-Aghdam, H.R., Williams, P.T. and Bartle, K.D., The role of exhaust pipe and incylinder deposits on diesel particulate composition, SAE Paper 921648, 1992.
12. Boone, E.F., Petterman, G.P. and Schetelich, Low sulfur fuel with lower ash lubricating oils – A new recipe for heavy duty diesels, SAE 922200, 1992.
13. Hercamp, R.D., Premature loss of oil consumption control in a heavy duty diesel engine, SAE Paper 831720, 1987.
14. Kim, Joong-Soo; Min, Byung-Soon; Lee, Doo-Soon; Oh, Dae-Yoon; Choi, Jae-Kwon, The characteristics of carbon deposit formation in piston top ring groove of gasoline and diesel engine, SAE 980526, 1998.
15. Bardasz, E.A.; Carrick, V.A.; George, H.F.; Graf, M.M.; Kornbrekke, R.E.; Pocinki, S.B.; Understanding soot mediated oil thickening through designed experimentation - Pa–t 4: Mack T-8 test, SAE 971693, 1997.
16. Van Dam, W.; Kleiser, W.M.; Lubricant related factors controlling oil consumption in diesel engines, SAE 952547, 1995.
17. Andrews, G.E., Li, H., Jones,M., Hall, J., Rahman, A.A. and Saykali, S., The influence of an oil recycler on lubricating oil quality with oil age for a bus using in service testing, SAE Paper 2000-01-0234, 2000.
18. Andrews, G.E., Hu Li, Hall, J., Rahman, A.A. and Saykali, S., The influence of an on line heated lubricating oil recycler on emissions from an IDI passenger car diesel as function of oil age, SAE Paper 2000-01-0232, 2000.
19. Harpster, M.J.; Matas, S.E.; Fry, J.H.; Litzinger, T.A., An experimental study of fuel composition and combustion chamber deposit effects on emissions from a spark ignition engine, SAE Paper 950740, 195.

20. McGeehan, J.A. and Fontana, B.J., Effect of soot on piston deposits and crankcase oils – infrared spectrometric technique for analysing soot, SAE Paper 901368, 1990.
21. K.Iwakata, Y.Onodera, K.Mihara and S.Ohkawa, "Nitro-oxidation of lubricating oil in heavy-duty diesel engine", SAE 932839
22. Bardasz, Ewa A.; Carrick, Virginia A.; George, Herman F.; Graf, Michelle M.; Kornbrekke, Ralph E.; Pocinki, Sara B., "Understanding soot mediated oil thickening through designed experimentation~Part 5: knowledge enhancement in the General Motors 6.5 L", SAE 972952
23. Bardasz, Ewa A.; Carrick, Virginia A.; George, Herman F.; Graf, Michelle M.; Kornbrekke, Ralph E.; Pocinki, Sara B., "Understanding soot mediated oil thickening through designed experimentation~Part 4: Mack T-8 test", SAE 971693
24. Parker,D.D. and Crooks,C.S., "Crankshaft bearing lubrication formation component manufacturer's perspective", IMechE Seminar on Automotive Lubricants: Recent Advances and Future Developments, S606, 1998.
25. York, M.E., "Extending engine life and reducing maintenance through the use of a mobile oil refiner", SAE831317
26. Byron Lefebvre, "Impact of electric mobile oil refiners on reducing engine and hydraulic equipment wear and eliminating environmentally dangerous waste oil", SAE 942032
27. Loftis, Ted S.; Lanius, Mike B., "A new method for combination full-flow and bypass filtration: venturi combo", SAE 972957
28. Stehouwer,D.M., "Effects of extended service intervals on filters in diesel engines", Proc. International Filtration Conference-The unknown Commodity, Southwest Research Institute, July 1996
29. Samways,A.L. and Cox,I.M., "A method for meaningfully evaluating the performance of a by-pass centrifugal oil cleaner", SAE 980872
30. Verdegan, Barry M.; Schwandt, Brian W.; Holm, Christopher E.; Fallon, Stephen L., "Protecting engines and the environment~A comparison of oil filtration alternatives", SAE 970551
31. Covitch, M.J., Humphrey, B.K. and Ripple, D.E., Oil thickening in the Mack T-7 engine test – fuel effects and the influence of lubricant additives on soot aggregation, SAE Paper 852126, 1985.
32. Ripple, D.E. and Guzauskas, J.F., Fuel sulphur effect on diesel engine lubrication, SAE Paper 902175, 1990.
33. Hartmann, J. High performance automotive fuels and fluids, Motor books International, p. 24, 1996.
34. Caines, A. and Haycock, H., Automotive Lubricants Reference Book, 1996.

CONTACT:
Professor G E Andrew, Dept. of Fuel and Energy, The University of Leeds, Leeds LS7 1ET, U.K.
Tel: 0044 113 2332493 Email: g.e.andrews@leeds.ac.uk

Dr Hu Li, Dept. of Fuel and Energy, The University of Leeds, Leeds LS7 1ET, U.K.
Tel: 0044 113 2440491 Email: fuelh@leeds.ac.uk

Mr James Hall, Top High (UK) Ltd, 918, Yeovil Road, Slough SL1 4JG , U.K
Tel: 0044 01753 573400 Email: tophigh@tophigh.co.uk

Greenhouse Gas Emissions: A Systems Analysis Approach

Konrad Saur, Kevin Brady and Andrea J. Russell
Five Winds International

ABSTRACT

The combustion of fossil fuels is increasing atmospheric concentrations of greenhouse gases and the weight of scientific evidence indicates that these gases are producing an enhanced green house effect that is altering the global climate. To address this challenge major industrialized countries have signed onto the Kyoto Protocol to the United Nations Framework Convention on Climate Change (UNFCCC). In response leading companies around the world are now investigating opportunities and evaluating the risks associated with their greenhouse gas (GHG) emissions. Creating a successful market for GHG credits will depend, in large part, on the development of credible measurement and verification protocols. Decision-makers need to be assured that an improvement option undertaken in a manufacturing plant does not result in upstream or downstream changes that will increase the overall release of GHGs. In this paper we argue that by using a systems analysis approach companies can have a broader perspective that will enable to not only measure their GHG baselines but also identify reduction opportunities and evaluate trade-offs. In addition, the approach fits into an environmental management system framework and is based on internationally accepted standards. A case example will be used to demonstrate this approach and to show the short and long term benefits of such a perspective.

INTRODUCTION

The World Meteorological Organization (WMO) announced that 1998 was "by far the warmest year since world-wide instrument records began 139 years ago." Global temperature has risen in the past 20 years faster than in any other 20-year period on record.

To address this situation the United Nations developed the Framework Convention on Climate Change (UNFCCC). The convention was originally open for signature at the 1992 United Nations Conference on Environment and Development (UNCED) and as of December 1999 it had been ratified by 181 member states. Subsequently the Kyoto Protocol to the convention was developed to strengthen the guidance to signatory bodies in regard to the policies and measures required to mitigate anthropogenic emissions of greenhouse gases. The protocol commits signatory countries or regions to specific emissions reduction targets. For example the US target is a 7% reduction, the European Community target is an 8% reduction and the Australian target is 8% increase over 1990 levels.

These emissions targets must be achieved by the period 2008-2012 and they will be calculated as an average over the five years. Cuts in the three most important gases - carbon dioxide (CO_2), methane (CH_4), and nitrous oxide (N_2O) - will be measured against a base year of 1990 (with exceptions for some countries with economies in transition). Cuts in three long-lived industrial gases - hydrofluorocarbons (HFCs), perfluorocarbons (PFCs), and sulphur hexafluoride (SF6) - can be measured against either a 1990 or 1995 baseline[1].

National governments and international agencies are now deep into the process of identifying policies and programs to meet their Kyoto reduction commitments. Solutions will likely come from a mix of regulatory programs, market based instruments, technology development and voluntary programs. Early indications are that governments will be developing policies and programs that favor less carbon intensive energy supply options and products. Consequently there is a renewed and growing interest in alternative energy technologies (e.g. fuel cells, solar, wind), cleaner fuels (e.g., biofuels, enhanced natural gas) cleaner electric power generation options (hydro and gas versus coal) and more carbon efficient products (hybrid vehicles).

Leading companies are trying to determine how to best reduce their GHG emissions and how to gain political and potentially economic credit for doing so. Effective reduction of GHG requires credible and accurate measurement approaches that help companies determine where to reduce the emissions they are responsible for. For example in order to gain credit for emissions reductions in an emissions trading system, companies must be able to prove that the reductions are real and credible. In addition, it is important that companies achieve greenhouse gas emissions

---

[1] UNFCCC Climate Change Information Kit. Chapter 21:The Kyoto Protocol.
http://www.unfccc.de/resource/iuckit/fact21.html

reductions without increasing other environmental emissions/impact in the process.

Companies can use a GHG systems analysis approach to identify a 1990 baseline for their emissions and to determine how and where to make changes to their current systems in order to reduce GHG emissions in a transparent and verifiable way. For a company, using internationally accepted approaches increases the credibility of emissions credits, while using a systems analysis approach decreases uncertainty, thereby increasing both the political and economic value of reductions.

MEASUREMENT TOOLS

In addressing global climate change it is imperative that the measurement tools utilized provide an accurate picture of where and how GHGs are being reduced. Because the point of release for GHGs is not relevant to the potential impact (i.e. a ton of CO_2 equivalents is equal to a ton of CO_2 equivalents no matter where it occurs) the measurement tool must be capable of assessing system wide implications of any action. For example, decision-makers need to be assured that an improvement option undertaken in a manufacturing plant does not result in upstream or downstream changes that will increase the overall release of GHGs. Such increases could arise from a large number of scenarios such as an upstream switch to more carbon intensive materials and processes or a design change that requires the product to be made from non-recyclable material. By tracking energy flows and GHG releases throughout all of the stages in the product life cycle LCA provides a system-wide perspective that enables decision-makers to see and evaluate these trade-offs.

It is also important that in the rush to reduce GHGs we do not aggravate other environmental concerns such as resource depletion, solid and hazardous waste generation, and the release of toxic substances which impact human health and ecological systems. Therefore the measurement tools utilized must be holistic and not focus on solely greenhouse gases. Life cycle based systems analysis meets this criteria as it not only tracks energy and energy and non energy related GHG releases, studies typically also track the consumption of other resources such as water and materials as well as multiple environmental releases (e.g. wastes, ozone depleting substances, toxic substances and common air pollutants). This holistic perspective helps the commissioner of the study understand the trade-offs inherent in any change to the system and it helps ensure that a reduction in GHG is not does not result in other adverse impacts such as increased release of toxic substances to the environment.

INTERNATIONAL STANDARDS - Another consideration in the choice of measurement tools is the need for globally accepted procedures and rules. Emissions Trading, Joint Implementation (JI) and the Clean Development Mechanism (CDM) involve partnerships among players in different countries, different sectors and in the case of the CDM countries at different stages of economic development (see Box 1). To have confidence as a "buyer" of emissions reductions one must have a means to evaluate the proposed project or reduction option and evaluate it against a credible internationally accepted standard. This confidence is required to ensure that the emission reduction is real and verifiable and to demonstrate to other stakeholders that the "buyer" has shown due diligence with respect to ensuring the emissions are real. To help increase confidence there is a need to develop standardized measurement and verification protocols.

Box 1.   Flexibility Mechanisms of the Kyoto Protocol

Clean Development Mechanism (CDM): The CDM provides a vehicle for emission reduction projects (investments) by countries listed in Annex 1 of the UNFCCC within non Annex 1 countries. Credits, for emissions reductions achieved through CDM projects will require some form of third party verification and they are called Certified Emission Reductions (CERs).

International Emissions Trading: A system of internationally tradable emissions permits based on national reduction targets and national inventories of sources and sinks for greenhouse gases. The quantity of permits issued to each Party (country or region e.g., the European Community) that accepts a cap on future GHG emissions is equal to that Party's assigned amount (AAs).

Joint Implementation (JI): A system of international cooperation designed to finance and implement cost-effective projects to reduce GHG emissions. It is a mechanism for cooperation only between Parties listed in Annex 1 of the UNFCCC. In enables the participating countries to transfer or acquire Emission Reduction Units (ERUs) from emission reduction or sink enhancement projects.

The International Organization for Standardization (ISO) has developed standards that can support this need.

The ISO 14040 series of Life Cycle Assessment standards developed by ISO Technical Committee 207 (TC207) are now completed. These standards are flexible and allow the commissioner of the study to design it to meet their own particular objectives (goal and scope of the study) using a systems analysis perspective. The standards provide direction on setting appropriate system boundaries, developing reliable data collection and handling procedures, evaluating and interpreting data and reporting in a transparent manner. This set of procedures and guidance offers an excellent starting point for the development of measurement protocols for GHGs. This is particularly true if the LCA standards are considered in conjunction with an environmental management system, which can provide a framework for setting objectives and targets related to GHG reduction.

ISO TC207 Climate Change Task Force (CCTF) has evaluated the potential relevance of the ISO standards to the issue of global climate change and they are now actively promoting the use of the standards by parties involved in the UNFCCC activities. With respect to LCA, the CCTF noted "in particular, ISO 14041, the Life Cycle Inventory (LCI) standard, can assist organizations in measuring greenhouse gas emissions and other environmental impacts. It may be used to establish a baseline of greenhouse gas emissions for a product system to benchmark environmental improvements or to evaluate alternatives. This can be done for the whole system but also broken down by unit process (e.g., electricity production or transportation). Specifically the CCTF report noted that the standard can be used to:

- Develop quantitative inventories of the greenhouse gas emissions associated with a product system.

- Develop quantitative inventories of the greenhouse gas emissions of the unit processes that make up a product system (e.g. electricity production, transportation).

- Provide data and information to identify which unit processes have the greatest use of energy and the greatest emissions.

- Identify energy efficiency improvement opportunities and other options to reduce greenhouse gas emissions.

- Provide data and information to evaluate the effect of new energy efficient technologies on the overall environmental profile (i.e. taking into considerations system-wide trade-offs).

The work of the CCTF indicates that systems based approaches are potentially very useful in helping organizations measure and manage their GHG emissions. In the next section we provide a specific example of how this is being accomplished, consistent with the international LCA standards developed by ISO.

CASE STUDY

To illustrate the systems analysis approach to GHG we are presenting a previous case study that features an LCI and Life Cycle Impact Assessment (LCIA) comparison of five different fender designs for an average compact class automobile in Germany. The goal of the case study is to determine the best material choice based on a systems analysis of GHG's released during the lifecycle of the fender. The five different product alternatives are steel sheet, primary aluminum sheet and three injection molded polymer blends; PP/EPDM, PPO/PA, PC/PBT. The technical specifications of all five fenders are identical; ensuring that the functional unit

applies to each and thus, that the fenders will be comparable. Table 1 below shows the materials and weights of the alternative fender designs.

Material	Thickness [mm]	Weight 2 Fenders [kg]
Steel	0.75	5.6
Aluminum	1.1	2.77
PP/EPDM T10	3.2	3.21
PPO/PA	3.2	3.35
PC/PBT	3.2	3.72

Table 1: Fender Design Properties

The study is based on production information for the year 1997 in Germany. All data and information used in the study are based on industry average. The life cycle of the fenders begins with the production and distribution of the raw materials. The study includes all transportation steps as well as the country or plant specific energy supply. The systems analysis includes the utilization phases of the different fender systems including the differences in fuel consumption due to their weight.

	Steel	Aluminum	Polymers
Material production	Coal and coke considered supplier specific (1995). Complete steel works inclusive milling and galvanizing balanced for suppliers (1995)	Alumina production and electrolysis (incl. energy supply) balanced site specific for all suppliers (1995/1996)	Polymer productions are averaged European industrial data (1993-96)
Processing	Press shop German average. Punching residues 100% recycling.	Press shop German average. Punching residues 100% recycling.	Material specific injection molding data averaged from several suppliers. Production scrap recycling considered
Recycling	Punching residues are recycled 100% in the converter.	Punching residues are recycled in the melting furnace, incl. salt slag recycling. Sec.-Al replaces primary material.	Recycled production residues replace virgin material.
Energy supply	Supplier specific; country grid mix or plant specific	Supplier specific; country grid mix or plant specific	Supplier specific; country grid mix or plant specific
Transport	Supplier specific	Supplier specific	Supplier specific
Data quality	Supplier specific (1995) and literature.	Supplier specific (1994).	Supplier specific (93-96) and literature.

Table 2: Data Origin and Boundary Conditions for Entire Study

ANALYSIS - The study reveals that the most interesting life cycle stage for the fenders is the use phase. The total weight of the vehicle is an extremely important parameter in determining the fuel consumption, and thus GHG emissions, of the vehicle. The weight of each part of the car has a share in the energy consumption during the utilization phase. Measurements and calculations performed by all automobile producers show that weight reductions can produce a range of 2.5% to 6% reduction in fuel consumption for each 10% in weight reduction. Table 2 shows the data origin and boundary conditions for the study.

The inventory results for the five fender designs are discussed below relative to energy demand and GHG emissions. The discussion is focused on selected resource and emissions parameters of interest to Climate Change. The study included an analysis of the entire inventory, with more than 30 resource parameters, 80 different air emissions and more than 60 waterborne emissions and different types of waste.

Figure 1 shows the energy consumption for the production phase of the different fenders. The main use of energy for this stage of the product lifecycle is the production of the raw materials used for each type of fender.

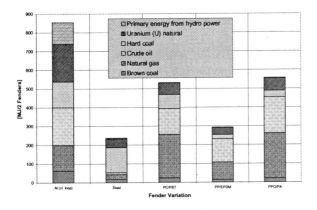

Figure 1: Energy Use During Production Stage

The aluminum fenders require the highest energy input of the five fender designs. This is a result of the electrolysis-process and the alumina production process. The steel fender has the lowest energy demand during the production stage. The energy demand of the PP/EPDM fender production is in the same range, whereas the more complex polymers PPO/PA and PC/PBT have a higher energy demand for material production.

Figure 2 shows the basic data and boundary conditions for the calculation of the utilization phase.

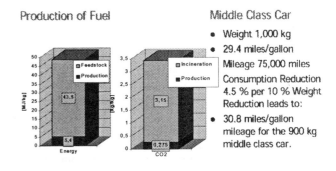

Figure 2: Boundary Conditions of the Fender Life Cycle Use Phase

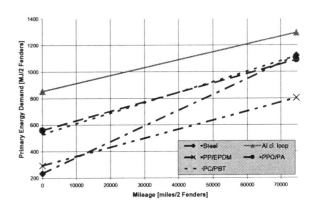

Figure 3: Energy Use During Production and Use Stages

Figure 3 shows the primary energy demand for the production and use stages of the fenders. The energy demand is displayed on the y-axis, dependent on the mileage on the x-axis. The values at 0 miles represent the primary energy demand needed for the material and part production, inclusive of all process steps.

The ascending gradients represent the different fuel consumption factors due to the varying weight of the fenders. The figure shows that the advantage of the steel design, which had the lowest energy consumption during production, is balanced by the higher fuel consumption during use. This is a result of the steel having the highest weight of the five fender designs.

The PP/EPDM fenders are the lightest design, which balances the higher energy consumption during production when compared with the steel fenders after mileage of approximately 12,000 miles. At 75,000 miles, the other polymer designs also balance energy consumption compared with the steel fenders. The aluminum fenders do not reach this break-even point within the mileage lifetime of the vehicle due to their high energy demand during the production stage.

Due to the combustion of fossil fuels, the CO2-Emissions into air in fig. 4 should, theoretically, display the same results as fig. 3. However, as a result of the feedstock energy that is stored in the polymer fenders, the CO2-Emissions of the polymer designs are relatively smaller than the metal fenders. Thus the CO2-Emissions of the PP/EPDM fenders are smallest from the start of the use phase. Also, the aluminum design reaches a break-even point when compared with the steel fenders. The reason for this is the high amount of hydro power used as an energy source during the production of the aluminum. This is contrary to steel production, which emits relatively high quantities of CO2 as a result of low efficiency and the use of hard coal.

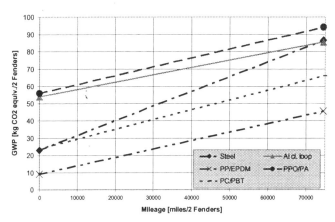

Figure 5a: GWP of Fender Designs – Production and Use

The life cycle inventory of the five different fenders shows, that the steel design has advantages in energy demand during production. However, when considering the use phase the overall energy consumption of the PP/EPDM fender is less due to the weight of the polymer design.

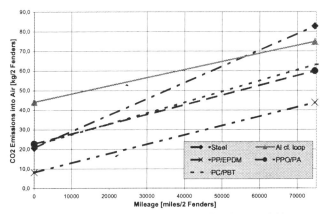

Figure 4. CO2 Emissions During Production and Use Stages

The contribution of each fender design to global warming potential (GWP) is visible in Figure 5a. These results are mainly influenced by the CO2-emissions due to energy demand, with one exception; the production of the PPO/PA fenders has the highest contribution to GWP due to high N2O-emissions. Second is the aluminum design as a result of the high CO2-emissions during aluminum production. The CF4 and C2F6-Emissions of the aluminum electrolysis make up approximately 14 % of the GWP of aluminum production. The other materials have a GWP corresponding to their CO2-emissions caused by the consumption of fossil fuel energy sources.

During the use phase, the CO2-Emissions of fuel consumption and emissions of methane caused by the oil mining for fuel production contribute to the GWP of each fender. As Fig. 5a shows, the PP/EPDM fenders have the smallest GWP within mileage of 75,000 miles. The aluminum design has an advantage in the use phase due to its low weight. Because of the high offset of its production, however, there is a late break even point at about 1,000,000 miles when compared with PP/EPDM.

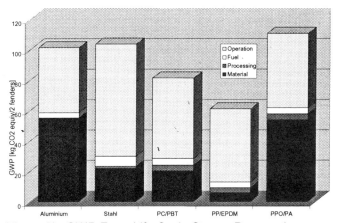

Figure 5b: GWP From Life Cycle Stages Perspective

DISCUSSION

The impact assessment shows that the GWP of the PP/EPDM design is the smallest even during the production of the fenders. The inventory shows that the use phase of automotive parts is dominated by the fuel consumption and related emissions such as CO2. Nevertheless, the impact assessment shows that some impacts are dominated by the production phase, for example the N2O-emissions of the PPO material production contribute to the high GWP of the PPO/PA-fender, which cannot be balanced within the use phase.

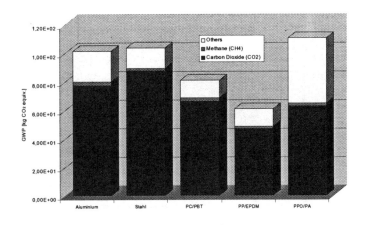

Figure 5c: GWP – Inventory Parameters Perspective

Figures 5a to 5c show clearly that a life cycle perspective offers opportunities for better understanding the systems perspective, thereby allowing for the identification of trade-offs.

This study highlights many important issues when considering greenhouse gas reductions in automotive applications:

- The results shown above show that specific emissions resulting from material production (e.g. N2O) and part processing are often of equal or even dominating importance (when compared with the use phase).

- During the use phase, a 'lightweight' part has the potential to balance higher energy consumption resulting from production.

- Lightweight materials, weight reduction and fuel savings are important topics for the future development of automotive parts and entire cars with respect to reducing greenhouse gas emissions during the life cycle.

- Improvement approaches focusing only on single aspects of the life cycle could lead to increases in greenhouse gas emissions in other stages.

- System boundaries and assumptions influence the results of life cycle considerations. They must be set carefully (i.e. if recycled material was used, Aluminum (recycled) would have been the most favorable design from a greenhouse gas perspective).

CONCLUSION

The analysis performed here has shown the value of using a life cycle based systems perspective when measuring the GWP of a product or system. By evaluating the use phase in addition to the traditional production phase of the lifecycle, several trade-offs and hidden impacts were discovered. For example, if one only examined the production stage, the obvious choice would be to use steel for the fender design, due to its low energy demand. Yet when examining the use phase, steel becomes unattractive, due to its weight. Also, the production energy used to produce the steel comes mostly from the burning of fossil fuels in this case, thus although its energy demand is lower than the other materials it has a higher contribution to GWP than the production of some other more energy intensive materials that use hydro or alternative power sources. As a company trying to reduce the amount of GHGs it contributes and gain emissions credits, this approach reduces uncertainty and clearly uncovers trade-offs, which may not have been uncovered using a traditional production only based perspective. If the case study company had chosen steel after only examining the production phase, they would be causing an increase in the amount of GHGs later in the product lifecycle.

Using a methodology to measure GHGs that is internationally recognized and has standards in place has definitive benefits. The systems analysis methodology helps a company's emissions credits gain more value, both in economic and political terms. It reduces uncertainty in study results due to its lifecycle perspective, thereby allowing companies to be more thorough in evaluating trade-offs during the decision-making process. The approach also fits into an environmental management system framework and can be verified using the ISO 14040 standards.

REFERENCES

1. ISO 14040: Environmental Management – Life Cycle Assessment – Principles and Framework.
2. ISO 14041: Environmental Management – Life Cycle Assessment – Inventory and Goal and Scope Definition.
3. FDIS 14042: Environmental management - Life cycle assessment - Life cycle impact assessment.
4. Chapter in Forthcoming Publication, United States EPA: Life Cycle Engineering, Dr. Konrad Saur

CONTACT

In Europe

Dr. Konrad Saur
Director
Five Winds International
k.saur @fivewinds.com

In North America

Kevin Brady
Director
Five Winds International
k.brady@fivewinds.com

About the Firm

Five Winds International is an international management consulting firm that assists organizations in improving environmental, social and economic performance. The company specializes in helping companies develop strategies, management systems, programs, tools and data needed to integrate environmental and social considerations into core business functions. These functions include strategy development, product design, supply chain management, sales and marketing, procurement and capital investments. The firm has offices in Germany, Canada and the United States and a network of contacts and associates across the globe. For more information please visit our website www.fivewinds.com.

2001-01-1081

Future Light-Duty Vehicles: Predicting their Fuel Consumption and Carbon-Reduction Potential

F. AuYeung, J. B. Heywood and A. Schafer
Massachusetts Institute of Technology

ABSTRACT

The transportation sector in the United States is a major contributor to global energy consumption and carbon dioxide emission. To assess the future potentials of different technologies in addressing these two issues, we used a family of simulation programs to predict fuel consumption for passenger cars in 2020. The selected technology combinations that have good market potential and could be in mass production include: advanced gasoline and diesel internal combustion engine vehicles with automatically-shifting clutched transmissions, gasoline, diesel, and compressed natural gas hybrid electric vehicles with continuously variable transmissions, direct hydrogen, gasoline and methanol reformer fuel cell hybrid electric vehicles with direct ratio drive, and battery electric vehicle with direct ratio drive. Using appropriately researched assumptions and input variables, calculations were performed to estimate the energy consumption and carbon dioxide emissions of the different technology combinations. Comparing the results for the vehicle driving cycle only, an evolutionary fuel consumption improvement of about 35 percent can be expected for the baseline gasoline car, given only market pressures and gradual regulatory requirements. With more research and investment in technology, an advanced gasoline engine car may further reduce fuel consumption by 12%, and a gasoline electric hybrid by 40%, as compared to the evolutionary car. Diesel versions of the advanced combustion and hybrid vehicles may be 10-15% better than their gasoline counterparts. Compressed natural gas hybrid vehicle may reduce fuel consumption by 3-4% but may reduce carbon dioxide emission by 25%. Meanwhile, a direct hydrogen fuel cell electric hybrid vehicle may have the greatest improvement over the baseline at 55%, but the gasoline and methanol reformers fuel cell versions appear very expensive and offer little benefit. Finally, aside from critical battery limitations, the electric vehicle is difficult to compare to other vehicles without taking into account the electricity generation process.

INTRODUCTION

The transportation sector is a substantial user of energy and a major source of anthropogenic carbon dioxide associated with global climate change. The United States, with less than 5% of the world's total population, consumes 25% of the world's current energy production and generates about 25% of the world's carbon emissions [1]. Within the U.S., where some 25% of the world's cars and trucks reside and operate [2], about a quarter of the national energy used goes to transportation [3].

To examine the potential for reducing this resource use and its associated emissions, the Massachusetts Institute of Technology Energy Laboratory has assessed the future of different potential passenger vehicle technologies, to compare their relative well-to-wheels energy consumption in a complete life-cycle analysis, and to identify possible barriers to implementation. This paper describes that part of the study dealing with the vehicle driving simulation and the vehicle tank-to-wheels fuel consumption results, focusing on the design criteria, vehicle performance, and ownership costs of future light-duty passenger cars for the U.S. market. We will describe the potential design options for future vehicles, our rationale for projecting technological advances, the key assumptions for the calculations, then report on the results and analyses.

APPROACH AND CONCEPTS EXAMINED

We have examined several potentially promising future powerplants and vehicle technology combinations using a computer simulation. This simulation "drives" the vehicle through a specified driving cycle, and calculates the fuel consumed and thus the carbon dioxide emissions produced. Inputs for the calculations are the vehicle driving resistances (mass or inertia, aerodynamic drag, and tire rolling friction) and the operating characteristics of each of the major propulsion system components (e.g. engine and transmission performance and efficiency for a standard internal combustion engine). These vehicle fuel consumption predictions are made for 2020, for technologies that could plausibly be in mass production at that time. Their estimated performance characteristics relative to today's performance include improvements that we judge could be implemented in production by 2020. However, the more sophisticated of these concepts, which could provide substantially improved fuel economy, are likely to be significantly more expensive. With the uncertainty in vehicle user response, our predictions indicate the fuel consumption and CO_2-reducing *potential* of various future propulsion systems and vehicle technologies, and do not express our judgments about either the desirability or the likelihood of these various technologies being in large scale production by 2020.

Energy-conserving technologies were chosen from a larger set of possible powertrain and vehicle developments as having the highest potential for reaching production and the market. Table 1 categorizes the combinations of propulsion system (power unit and transmission) and fuels examined into three families: mechanical, hybrid, and electrical.

FAMILY	TRANSMISSION	POWER UNIT	FUEL
Mechanical	Auto-Clutch	Spark Ignition ICE	Gasoline
		Compression Ignition ICE	Diesel
Dual	Continuously Variable	ICE with Batteries and Electric Motor	Gasoline, Diesel, Natural Gas
Electrical	Single Ratio	Fuel Cell (with reformer for gasoline, methanol)	Gasoline, Methanol, Hydrogen
		Battery	Electricity

Table 1. Powerplant and fuel combinations examined

An important issue is the relevant baseline for comparisons. We have selected an evolving mid-size US passenger car: i.e., a steadily improving gasoline-fueled spark-ignition engine, with a more efficient conventional technology transmission and low cost vehicle weight and drag reductions. The baseline technology improvements are based on historical and current technology trends [4] projected to 2020. They represent the likely average passenger car technology in 2020 that will only incur those extra costs necessary to keep up with the market.

A second issue is the performance and operating characteristics of these various future vehicle and powerplant combinations. Ideally, each combination should provide the same (or closely comparable) acceleration, driveability, driving range, refueling ease, interior driver and passenger space, trunk storage space, and meet the applicable safety and pollutant emissions standards. Only some of these attributes can be dealt with quantitatively now.

We attempt to hold these attributes essentially constant as we project the impact of technology changes on the average car. Whether or not customers will maintain their expectations roughly constant is unclear. Over the past two decades, the performance of specific vehicle models has increased, as has weight and energy consuming on-board features. New vehicle concepts have been introduced into the fleet (e.g. sport utility vehicles); these are usually larger and heavier than the average passenger car and hence are higher energy consuming. Especially critical here are future customer attitudes to real or perceived safety issues that reductions in a given model vehicle weight may create.

In this study, to maintain constant performance, all propulsion system and vehicle combinations are adjusted to provide the same ratio of maximum power to total vehicle mass, and provide at least 600 km driving range, except for the special case of the pure electric vehicle, whose battery constraints will be discussed

later. The vehicle size is roughly constant. Driveability issues (e.g. ease of start up, driving smoothness, transient response for rapid accelerations, hill climbing, and load carrying/pulling capacity) have not yet been assessed quantitatively. The emission levels projected to 2020 with these various technologies also have not yet been quantified. We assume that the strictest current emissions standards (California LEV II, EPA Tier II) for 2004 to 2008 may be further reduced in the following decade, but that these levels can potentially be met by improved exhaust gas treatment technology for internal combustion engines, and are within expectations for fuel-cell systems. This assumption is least certain for the diesel ICE.

Third, in simulations to predict fuel consumption and carbon emission, there is an obvious sensitivity to assumptions concerning the performance characteristics of key vehicle and propulsion system components. This is especially challenging for new technologies that are still in their development stage, such as fuel cells and their associated fuel reformers. There is substantial uncertainty in estimating their future performance. Optimism regarding technological progress must be balanced against the compromises that the cost constraints and robustness requirements of the mass production car market demand.

Finally, we note that the full potential and impact of these various technologies can only be assessed by taking into account both the fuel and vehicle cycles. For example, the hydrogen fuel cell vehicle and the battery-electric vehicle are both classified as zero-emission when only the vehicle cycle is considered. However, the inclusion of the fuel cycle, i.e. how the hydrogen and electricity are generated, reveals that although the technologies may potentially be clean, zero-emission from the total system is unlikely.

SIMULATION MODEL STRUCTURE

A family of Matlab Simulink simulation programs was used to estimate fuel consumption to compare various vehicles with different propulsion systems,. Originally developed by Guzzella and Amstutz at the Eidgenössiche Technische Hochschule (ETH, Zurich [5]), these programs back-calculate the fuel consumed by the propulsion system by driving the vehicle through a specified cycle. Such simulations require performance models for each major propulsion system component as well as for each vehicle driving resistance. The component simulations used, which are updated and expanded versions of the ETH simulations, are best characterized as aggregate, quasi-steady, engineering models which quantify component performance in sufficient detail to be reasonably accurate, but avoid excessive detail which would be difficult to justify for predictions relevant to 2020.

The first critical component of the simulation is the driving cycle on which all the vehicle calculations are based. For this study, the Federal Test Procedure (FTP) urban (city) and highway driving cycles are used. These cycles are the ones used by the Environmental Protection Agency (EPA) to measure the emissions and fuel consumption of vehicles sold in the U.S, reported each year in the EPA Fuel Economy Guide [6], after multiplying by an empirically determined factor (0.90 for

156

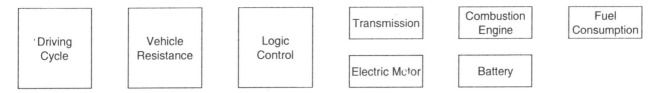

Figure 1. Gasoline and diesel performance maps

the city cycle and 0.78 for the highway cycle) to take into account additional real-life driving factors. The results presented in this report have <u>not</u> been multiplied by these empirical factors. A combined fuel consumption is calculated with 55% city and 45% highway by distance.

Our results will obviously pertain to this choice of driving cycle. Real world driving will give different (likely worse) fuel consumption. Also, the fuel consumption of different technology vehicles with different driving cycles will not have the same relative relationship to each other. However, the FTP is the standard US driving cycle.

The calculation starts with the chosen driving cycle. From the time vs. velocity inputs, vehicle acceleration is calculated. This information is then used to calculate the instantaneous power needed to operate the vehicle, by

adding aerodynamic drag, tire rolling resistance, and vehicle inertial forces. From this required total power, the torque needed to drive the wheels is calculated; this is transferred back to an engine through a transmission and/or coupled directly to a motor. For hybrids, a simple control logic determines the power split between powerplants. During idling and low-power operation, the batteries supply the necessary power. Over a certain threshold, the on-board power generator (combustion engine or fuel cell) turns on to drive the vehicle; extra power is used to recharge the batteries if they are below a set state of charge. When the power required exceeds the maximum capabilities, the batteries supplements peak engine output. Additional powerplant inefficiencies and auxiliary loadings are included, and then the total energy consumed per unit distance traveled is calculated. An example of the simulation logic for an ICE/electric hybrid vehicle is shown in Figure 1.

	current	baseline	advanced	advanced	advanced	advanced	advanced	advanced	advanced	advanced	advanced
	SI ICE	SI ICE	SI ICE	CI ICE	SI Hybrid	CI Hybrid	SI Hybrid	FC Hybrid	FC Hybrid	FC Hybrid	Electric
	gasoline	gasoline	gasoline	diesel	gasoline	diesel	CNG	gasoline	methanol	hydrogen	electricity
	auto	auto-clutch	auto-clutch	auto-clutch	CVT	CVT	CVT	direct	direct	direct	direct
Body	383	326	249	249	249	249	249	249	249	249	249
Glazing	35	33	33	33	33	33	33	33	33	33	33
Chassis	273	229	216	219	216	219	216	275	259	244	243
Propulsion System	392	263	252	303	267	302	283	536	475	416	414
Engine	164	103	95	149	64	99	67	0	0	0	0
Electric Motor					19	20	20	73	69	66	66
Fuel Cell System								351	278	193	0
Battery	12	12	12	12	36	37	37	46	43	41	328
Transmission	90	50	50	50	50	50	50	20	20	20	20
Liquids and Storage	64	45	42	39	34	31	46	33	53	84	0
Other	62	53	53	53	64	64	64	14	13	12	0
Interior&Exterior	195	214	214	214	214	214	214	194	194	194	194
Other	44	44	44	44	44	44	44	44	44	44	44
TOTAL VEHICLE	1322	1108	1007	1062	1023	1060	1039	1330	1253	1179	1176

Table 2: Mass Distribution in kilograms by component for all vehicles examined.

COMPONENT DETAILS

The vehicles examined in this study are designed to be functional equivalents of today's average passenger car: a mid-sized family sedan such as the Toyota Camry. For the customer, this means characteristics such as vehicle performance are maintained in future vehicles. All vehicles are designed to have an equal peak power to mass ratio of 75 W/kg, which is approximately today's value, and roughly equalizes vehicle performances.

Vehicle Body

The baseline reference vehicle is projected to incorporate evolutionary technology changes required to compete in the 2020 automotive market, while all other vehicles studied in this report undergo advanced body changes (including more radical and expensive new technologies to reduce vehicle mass and drag) that would be more appropriately matched to the advanced propulsion systems involved. Our analysis of vehicle mass projections and distributions are presented in Table 2.

The total mass is subdivided into subsystems for comparison: chassis and body, propulsion system, battery, fuel, and occupant. The chassis and body system mass include everything for an un-powered free-rolling vehicle, including the fuel system (but without the fuel) as well as all structural reinforcement for extra component masses on the vehicle. The propulsion system mass include the engine, electric motor, fuel cell system and reformer system, all scaled according to the desired power output, as well as the transmission. The battery pack size is determined by the maximum power required by the electric motor in a particular vehicle, resulting directly in a specific battery mass and volume. This then limits the amount of battery energy available. The fuel mass is the amount of fuel needed to achieve a range of 600 km in the combined cycle. An occupant mass is then added to the total raw vehicle mass; it is the estimated average load for a vehicle, held constant for all vehicles in this study at 136 kg (300 lbs), the mass of 1.5 adults and cargo.

Other key simulation variables include: vehicle aerodynamic drag coefficient C_d, cross-sectional area A_x, tire rolling resistance coefficient C_{rr}, transmission efficiency η_{trans}, and auxiliary load P_{aux}. Compounding effects also play an important, self-reinforcing role in vehicle mass reduction. By reducing body mass and lowering vehicle resistances, a smaller and lighter power unit can be chosen to meet the predetermined performance criteria, thus reducing the structural mass and inertial and rolling resistances even further. All technologies benefit from compounding effects, some more than others.

Gasoline/Diesel Engines

The performance characteristics of both gasoline and diesel internal combustion engines are well documented. Historical improvement trends, combined with an assessment of likely practical technologies available over the next two decades, are used to predict the performance of these two engines in year 2020.

A typical maximum engine torque curve was constructed for a 1.6 L gasoline engine and a 1.7 L turbo diesel engine. These torque-rpm curves can be scaled over a range of engine displacements, and define the performance of engines today. To project forward, historical trends correlating the ratio of gasoline engine power to displaced volume against year show a nearly linear improvement of about 0.5% per year [4]. Future technological improvements such as increasing use of variable valve timing, direct injection, and reduced friction are expected to continue this trend. Hence for 2020, the wide-open-throttle (WOT) torque for these engines is increased by 10% overall. Gasoline engines are expected to operate and generate peak power at a higher maximum speed with these and similar advancements. Thus, an extra cumulative 1% increase was added at each 500 rpm interval as engine speed increases for a 20% increase at maximum power. Engine characteristics are shown in Figure 2.

Internal combustion engine efficiency maps were modeled using a constant indicated energy conversion efficiency (fraction of fuel chemical energy transferred to the engine's pistons as work) and a constant friction mean effective pressure (total engine friction divided by displaced cylinder volume) [7]. This simple method is correct in aggregate but does not take into account the effect of engine speed on efficiency. However, over the normal engine speed range, this assumption is adequate

for predicting engine efficiency. The brake or useable engine output is obtained from the relation:

$$bmep = imep - fmep \qquad (1)$$

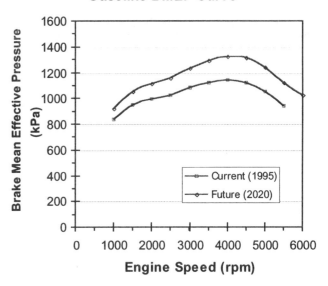

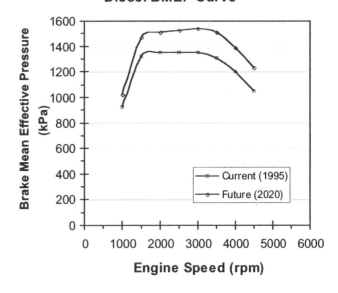

Figure 2. Gasoline and diesel performance maps

where bmep is the brake mean effective pressure (work produced per engine cycle divided by displaced volume). The indicated mean effective pressure is obtained from the indicated efficiency:

$$imep = \eta_i(m_f Q_{HV}/V_d) \qquad (2)$$

where the m_f is the fuel mass per cycle, Q_{HV} is the lower heating value of the fuel, and V_d is the total cylinder displaced volume.

Thus, the brake mean effective pressure used to determine engine torque (by scaling with displaced volume) is obtained from the indicated performance, offset by the friction of the engine. Based on projected technological improvements, the indicated efficiency in 2020 is assumed to increase by 7.5% to $\eta_i = 0.41$ for gasoline engines, $\eta_i = 0.44$ for CNG, and $\eta_i = 0.52$ for diesels. Engine friction is expected to decrease by 25% to fmep = 124 kPa for gasoline and CNG engines and by 15% to fmep = 153 kPa for diesel engines.

An internal Energy Laboratory study shows that the 1996 median new passenger car was a mid-sized sedan with averages of: base price US$ 18,124, base mass 1333 kg, and fuel consumption of 11.0 L/100km city and 7.9 L/100km highway. These characteristics closely match the selected current vehicle used in this study.

Battery Electric

Data are available to quantify the efficiency of pure electric vehicles, although its history is brief and uneven, [8] based on the extensive development though poor sales record of recent pure electric vehicles produced.

Since electric motors have been in service for many applications and have been tuned to optimize performance, a motor peak torque and power curve based on today's electric motor, as shown in Figure 3, can be used for the future as well.

A motor efficiency map (10×21 array) based on motor speed and torque output is used to model motor efficiency, while the power inverter is assumed to have a constant efficiency of 94%. Together with the modeled single gear ratio transmission loss, the total electric motor system efficiency is about 80% over the combined driving cycle. An additional 15% loss is added in turn-around operation when the motor is used in regenerative braking to convert mechanical work to electricity. For electric vehicles, a charging efficiency of 85% from an station or outlet is included in the vehicle cycle; this is equivalent to a user spilling gasoline at the pump.

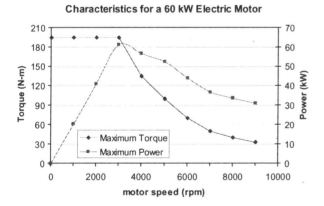

Characteristics for a 60 kW Electric Motor

Figure 3. Electric motor performance

Although other technologies are being developed, nickel metal hydride (NiMH) batteries are the technology of choice for automotive applications today, both for hybrids and electric vehicles. EV batteries currently have a specific energy of about 70 Wh/kg and a specific power of about 150 W/kg [9]. For 2020, it is assumed that battery performance will improve, especially the specific energy, and that battery performance will be close to meeting the Advanced Battery Consortium's (ABC) commercial goals of 150 Wh/kg and 300 W/kg [10], which is desirable to produce acceptable EV performance. Although NiMH cannot reach this potential, another technology such as the lithium-ion battery, may; its specific energy is significantly higher than that of the NiMH battery technology.

Batteries are not designed to be fully discharged, which would shortened their lifetime and decrease their capacity. Also, topping off the battery at high state of charge is not efficient given the internal resistance of the batteries. Hence, cycled battery applications tend to operate usually within the 20-80% SOC.

For the pure electric vehicle, battery performance density constraints are of great concern. In addition to providing the power needed for peak motor power, extra batteries may be added to increase the energy storage capacity and hence extend vehicle range. However, extra batteries also add to the vehicle mass and thus require increased motor power, additional structural support, and more batteries to maintain performance, generating an undesirable compounding effect. Given this constraint, the battery pack is selected based only on its power capacity. Also, the battery volume must also be considered because of its possible intrusion into the interior space.

Internal Combustion Engine Electric Hybrid

Data are becoming available for ICE-electric hybrid vehicles, especially for gasoline hybrids with two limited production versions already in the market. With several different types of feasible hybrid configurations, and different drivetrain arrangement within each configuration, the Toyota Prius version of parallel, duel-mode, CVT configuration was selected and slightly modified for the model.

Starting with the most basic distinguishing characteristic, there are series and parallel hybrids. A series hybrid drives the wheels only through the electric motor with the combustion engine generating electricity, whereas a parallel hybrid system powers the wheels directly with both the combustion engine and electric motor.

Within the parallel hybrid family, there is a further separation between dual-mode and power-assist, and between road-coupled and wheel-coupled configurations. A duel-mode drivetrain allows vehicle operation with just the engine, or just the motor, or with both, whereas a power-assist drivetrain always draws primary power out of the engine with the electric motor serving only to supplement high loads. Meanwhile, a road-coupled drivetrain has the two power sources, unconnected, driving different wheels, whereas a wheel-coupled drivetrain combines the engine and motor before transferring power to the wheels.

Within the parallel, dual-mode, wheel-coupled family, the electric motor can contribute power before or after the geared transmission for the combustion engine. The

Toyota Prius uses a planetary gear setup to couple the engine and the motor prior to the continuously variable transmission. In our study, the motor power bypasses the combustion engine/CVT combination, and drives the wheels directly through a single-speed gear ratio reduction for internal consistency with the pure electric drive vehicles and for improved efficiency.

Controlling the power balance between the combustion engine and electric motor is dependent on many factors, such as driver requirements, power demand, vehicle speed, and battery state of charge. Many options exist and could be very sophisticated. For the simulation, a simplified control model is used. During low power situations, only the electric motor is in operation, thus eliminating engine idling and the less efficient and more polluting modes of operation for combustion engines. Above a preset threshold, the vehicle will be driven only by the combustion engine, except at the higher loads, such as during hard acceleration or hill climbing, when the electric motor serves as a load-leveler and provides the necessary additional power to add to the engine's maximum output.

For hybrid systems, it is the batteries specific power that is critical, since discharged batteries can be recharged during ICE operation. HEV batteries currently have a specific power of about 420 W/kg and a specific power of about 65 W/kg [Ovonics]. For 2020, it is assumed that battery performance will improve, especially in specific power, and will reach a conservative goal of 800 W/kg and 50 Whr/kg [11]. Again, lithium-ion technology may likely reach and surpass this goal.

While all technologies are held to the same peak power to mass ratio, hybrid technologies have the extra factor in balancing power contribution between the engine and the motor. Having performed a series of varying power combinations, we find a difference in energy consumption of roughly 10%, after taking into account the battery state of charge and summing it with the fuel consumption to acquire an equivalent total energy use. Arguments for more engine or more motor power must be carefully weighed. On one hand, a larger engine means smaller battery/motor mass and better highway operation, when the ICE is more efficient; on the other hand, a larger motor means more effective regenerative braking energy capture and better dual-mode operation, when the electric motor is preferred in a city setting. In the end, motor power is chosen at 30% of total available power for the advanced ICE parallel hybrids.

Note that maintaining an adequate charge in the battery is a reasonable expectation for normal urban driving. For driving that requires high power over extended periods of time (such as long hill climbing or towing at high speeds), however, the battery charge may be depleted and the total system power will then be reduced. Conventional ICE vehicles do not suffer this penalty.

It is important to note that the continuously variable transmission also help reduce fuel consumption as compared to the advanced vehicles. Because of the hybrid mode, the combustion engine does not idle or operate below 2 kW. In addition, the motor allows for modest regenerative braking, recovering some of the energy that would otherwise be lost to heat. The CVT also improves the operating conditions by using high-efficiency regions of the engine, thus reducing energy consumption.

Fuel Cell Electric Hybrid

Data exist only for prototype fuel cell vehicles, and many details about component performance are unavailable. Also, significant fuel cell system technology improvements are occurring in stack size and weight for a given power, fuel storage methods, reformer performance, and cost. Modeling future production fuel cell systems that currently exist only in prototype form is speculative and uncertain, although overall system component efficiencies can be plausibly estimated.

			2020	2020
	Date		advanced	advanced
	Technology		FC Hybrid	Fuel Cell
	Propulsion System		hydrogen	hydrogen
	Fuel		direct	direct
	Transmission			
	VARIABLE	battery units		none
Mass	Body & Chassis Mass	kg	763	781
	Propulsion System Mass	kg	371	479
	Battery System Mass	kg	41	0
	Fuel Mass	kg	4.0	4.0
	Occupant Mass	kg	136	136
	Total Mass	**kg**	**1314**	**1399**
Power	Power:Weight Ratio	W/kg	**75.0**	**75.0**
	Max Motor Power	kW	98.5	105.0
	Peak Stack Power	kW	65.7	105.0
	RESULTS			
City	Fuel Energy Use	MJ/km	0.904	1.062
	Battery Status	MJ/km	0.000	
	Combined Energy Use	**MJ/km**	**0.905**	**1.062**
	Range (fuel only)	km	532	453
	Tank-to-Wheel Efficiency	%	36.2%	32.6%
Highway	Fuel Energy Use	MJ/km	0.698	0.716
	Battery Status	MJ/km	-0.009	
	Combined Energy Use	**MJ/km**	**0.684**	**0.716**
	Range (fuel only)	km	689	672
	Tank-to-Wheel Efficiency	%	35.8%	35.5%
combined	**Equivalent Energy Use**	**MJ/km**	**0.805**	**0.906**
	Gasoline Eq. Consumption	**L/100km**	**2.50**	**2.81**
	Gasoline Eq. Economy	**mpg**	**94.1**	**83.6**
	Cycle Carbon Emission	**g C /km**	**0.0**	**0.0**

Table 3. Hydrogen fuel cell hybridization

Unlike the combustion engine hybrid, the fuel cell battery hybrid operates with a pure electric drivetrain, with the fuel cell generating electricity that powers the electric motor and accessories, recharges the batteries, or both. (For fair comparison with the ICE hybrids, FC hybrids also utilize a 30% battery power ratio). A prior study [12] demonstrates that hybridization of fuel cell vehicles helps conserve fuel, a savings of 11% from the calculations using the model in this study, as shown in Table 3. Hybridization is also preferred and may be necessary for reformer fuel cell systems to eliminate the lag time of reformer warm-up and response to driver demand. The power logic control operates in a similar manner to that of the combustion hybrid.

The fuel cell system efficiency is based on modeling by Directed Technologies [13]. First, the power to efficiency curve is scaled to the stack size and to yield the gross power output. Then, 15% of the generated power is diverted to run the needed fuel cell systems. The resulting power vs. efficiency plot is shown in Figure 4.

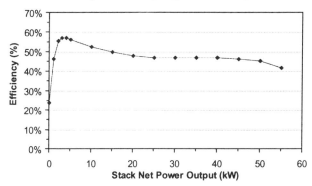

Figure 4. Fuel cell power vs. efficiency

An additional fuel cell system loss is taken into account for fuel reformer vehicles, where reduced hydrogen flow concentration results in poorer stack performance and incomplete hydrogen utilization. The methanol reformer generates a stream with 75% hydrogen, with a 10% reduction in fuel cell power; the gasoline reformer generates a stream with 40% hydrogen, with a 21.5% reduction in fuel cell power [13]. Because of the open flow of the input stream, both fuel cells have a hydrogen utilization rate of 85%. All numbers from Directed Technologies are taken as an average of their best and probable cases.

On-board reformer technologies are not well-developed and are the center of much research attention, making any prediction difficult. For the simulation, a lumped reformer efficiency is used: 82% for the methanol steam reformer and 72.5% for the gasoline partial oxidation reformer [13]. It is assumed that the batteries or a hydrogen reservoir will compensate for the lag time in power response.

VEHICLE SIMULATION RESULTS

We have first verified our simulation models on a set of current production and prototype vehicles: the Toyota Camry (4-cylinder manual and auto, and 6-cylinder automatic), the 1990 Audi 100 turbo diesel (5-cylinder manual), the Toyota Prius (4-cylinder CVT hybrid), the Ford P2000 prototype hydrogen fuel cell vehicle, and the GM EV1 (NiMH batteries) limited production electric vehicle. The urban and highway fuel economies are compared in Table 4. While not all input details for these vehicles are available, the results show reasonable agreement with Federal Test Procedure or company published data. The Partnership for a New Generation of Vehicles (PNGV) has also produced promising prototypes that include diesel hybrids and fuel cell vehicles that have achieved impressively low energy consumption.

Then, based on the described component details and assumptions, the vehicle simulations were performed, with the results and description shown in Tables 5 and 6.

The table summarizes the major component input variables, and component and vehicle results from the vehicle simulation calculations. The US FTP urban (city) and highway driving cycles were used; ten different vehicle and propulsion systems were examined. The first column (on the left) in Table 5 is a current average passenger car (note again that the EPA empirical factor of 0.90 for city and 0.78 for highway fuel consumption are not used for the results); the second column is the evolving baseline average car projected out to 2020.

The advanced technology vehicles (in 2020) are then arranged in four groups: internal combustion engine vehicles, internal combustion engine/battery hybrids, fuel cell hybrids, and electric vehicle. (Note that outlet charging is not included for the battery electric vehicle total energy consumption.) All these advanced technology vehicles have reduced vehicle resistances (mass, aerodynamics drag, tire resistance) compared with the 2020 baseline vehicle.

MODEL	*numbers in mpg gasoline equivalent* power unit	trans.	Published/Reported City	Highway	Unadjusted/Actual City	Highway	Simulation Result City	Highway	Percent Difference City	Highway
Toyota Camry	4-cyl gasoline	manual	23	31	25	39	28.3	39.1	13%	0%
Toyota Camry	4-cyl gasoline	auto	21	27	23	35	24.1	35.9	5%	2%
Toyota Camry	6-cyl gasoline	auto	20	29	23	37	22.6	32.2	-2%	-13%
Audi 100	5-cyl diesel	manual	33.1	41-56	37		37.9	53.0	3%	<10%
Toyota Prius	gasoline hybrid	CVT	lower 50's	lower 40's			39.8	46.4	<15%	<5%
Ford P2000	hybrid fuel cell	direct			56	80	55.6	69.9	-2%	-13%
GM EV1	battery electric	direct			100	113	93.2	120.8	-6%	7%

Table 4. Comparison of fuel economy results with existing data

The results at the bottom of Table 5 show energy use, fuel consumption/economy (including battery SOC in equivalent fuel use), range, overall vehicle energy efficiency (tank to wheel) for the urban and highway driving cycles, and the standard 55% urban 45% highway combined energy/fuel consumption average, and CO_2 emissions on grams carbon per average vehicle km traveled. Combined fuel economy and consumption are expressed in gasoline equivalent of the energy used. The calculated ranges of each of these vehicles are closely comparable (about 600 km) except for the EV, which depends strongly on the assumed battery characteristics.

Date		1996	2020	2020	2020	2020	2020	2020	2020	2020	2020	2020
Technology		current	baseline	advanced	advanced	advanced	advanced	advanced	advanced	advanced	advanced	advanced
Propulsion System		SI ICE	SI ICE	SI ICE	CI ICE	SI Hybrid	CI Hybrid	SI Hybrid	FC Hybrid	FC Hybrid	FC Hybrid	Electric
Fuel		gasoline	gasoline	gasoline	diesel	gasoline	diesel	CNG	gasoline	methanol	hydrogen	electricity
Transmission		auto	auto-clutch	auto-clutch	auto-clutch	CVT	CVT	CVT	direct	direct	direct	direct
VARIABLE	units											
Mass												
Body & Chassis Mass	kg	930	845	756	759	756	759	756	794	778	763	763
Propulsion System Mass	kg	340	226	217	271	216	251	235	465	390	371	86
Battery Mass	kg	12	12	12	12	36.0	37.2	36.6	45.5	43.0	41.0	328.0
Maximum Fuel Mass	kg	40.2	24.7	21.8	20.2	15.8	13.9	13.0	25.2	42.0	4.0	0.0
Occupant Mass (300 lbs.)	kg	136	136	136	136	136	136	136	136	136	136	136
Total Mass (2/3 tank)	**kg**	1444	1236	1136	1191	1154	1192	1172	1458	1375	1314	1312
Vehicle												
Rolling Resistance Coeff.	---	0.009	0.008	0.006	0.006	0.006	0.006	0.006	0.006	0.006	0.006	0.0060
Drag Coefficient	---	0.33	0.27	0.22	0.22	0.22	0.22	0.22	0.22	0.22	0.22	0.22
Frontal Area	m2	2.0	1.8	1.8	1.8	1.8	1.8	1.8	1.8	1.8	1.8	1.8
Auxiliary Power	W	700	1000	1000	1000	1000	1000	1000	1000	1000	1000	1000
Power:Weight Ratio	**W/kg**	76.0	75.0	75.0	75.0	75.0	75.0	75.0	75.0	75.0	75.0	75.0
Engine												
Engine Displacement	cm3	2500	1790	1645	1875	1114	1284	1360				
Transmission Efficiency	---	0.7-0.8	0.88	0.88	0.88	0.88	0.88	0.88				
Indicated Efficiency	---	0.38	0.41	0.41	0.51	0.41	0.51	0.44				
Frictional ME Pressure	kPa	165	124	124	154	124	153	124				
Max Engine Power	**kW**	109.7	92.7	85.2	89.4	57.7	59.6	58.6				
Motor												
Hybrid Threshold	kW					2.0	2.0	2.0	3.3	3.0	3.1	
Gear Efficiency	---					0.66	0.65	0.66	0.93	0.93	0.93	0.95
Electric Motor Efficiency	---					0.76	0.76	0.76	0.80	0.80	0.80	0.82
Max Motor Power	**kW**					28.8	29.8	29.3	109.3	103.1	98.5	98.4
Fuel Cell												
H2 Flow Concentration	%								40%	75%	100%	
Fuel Cell System Efficiency	---								0.41	0.47	0.52	
Reformer & Utilization Eff.	---								0.62	0.70		
Peak Stack Power	kW								72.9	68.7	65.7	
Fuel												
Lower Heating Value	MJ/kg	43.7	43.7	43.7	41.7	43.7	41.7	50	43.7	20.1	120.2	
Fuel Density	kg/L	0.737	0.737	0.737	0.856	0.737	0.856	0.16	0.737	0.792		
Battery												
Battery Discharge Efficiency	---					0.95	0.95	0.95	0.95	0.95	0.95	0.95
Specific Energy	Wh/kg					50	50	50	50	50	50	150
Specific Power	W/kg					800	800	800	800	800	800	300
RESULTS												
City												
Fuel Energy Use	MJ/km	3.195	1.997	1.786	1.582	1.200	1.029	1.153	2.038	1.513	0.904	
Battery Status	MJ/km					0.002	0.000	0.002	0.000	0.001	0.000	0.579
Combined Energy Use	**MJ/km**	3.195	1.997	1.786	1.582	1.209	1.029	1.160	2.038	1.517	0.905	0.579
Combined Fuel Consumption	L/100km	9.92	6.20	5.55	4.43	3.75	2.88	14.50	6.33	9.53		
Combined Fuel Economy	mpg	23.7	37.9	42.4	53.1	62.7	81.6		37.2	24.7		
Range (fuel only)	km	550	541	533	532	575	563	564	540	558	532	360
Tank-to-Wheel Efficiency	%	13.0%	16.9%	16.2%	19.0%	26.4%	31.8%	27.8%	17.6%	22.5%	36.2%	61.5%
Highway												
Fuel Energy Use	MJ/km	2.152	1.454	1.246	1.070	0.919	0.807	0.895	1.520	1.138	0.698	
Battery Status	MJ/km					-0.004	-0.005	-0.005	-0.009	-0.009	-0.009	0.422
Combined Energy Use	**MJ/km**	2.152	1.454	1.246	1.070	0.900	0.788	0.876	1.489	1.107	0.684	0.422
Combined Fuel Consumption	L/100km	6.68	4.51	3.87	3.00	2.79	2.21	10.95	4.62	6.95		
Combined Fuel Economy	mpg	35.2	52.1	60.8	78.5	84.2	106.5		50.9	33.8		
Range (fuel only)	km	816	743	765	787	751	719	726	724	742	689	494
Tank-to-Wheel Efficiency	%	17.1%	19.4%	18.1%	21.7%	25.7%	29.8%	26.6%	17.5%	22.7%	35.8%	58.8%
combined												
Equivalent Energy Use	MJ/km	2.726	1.753	1.543	1.352	1.070	0.921	1.032	1.791	1.332	0.805	0.508
Gasoline Eq. Consumption	L/100km	8.46	5.44	4.79	4.20	3.32	2.86	3.20	5.56	4.14	2.50	1.58
Gasoline Eq. Economy	mpg	27.8	43.2	49.1	56.0	70.8	82.3	73.4	42.3	56.9	94.1	149.1
Cycle Carbon Emission	g C /km	53.3	34.3	30.2	28.2	20.9	19.2	15.5	35.0	24.9	0.0	0.0

Table 5. Summary results for test vehicles. See Table 6 for additional information.

	VARIABLE	units	
Mass	Body & Chassis Mass	kg	see vehicle mass distribution
	Propulsion System Mass	kg	see vehicle mass distribution
	Battery Mass	kg	see vehicle mass distribution
	Maximum Fuel Mass	kg	except for electric vehicle, fuel is scaled for ~600km range.
	Occupant Mass (300 lbs.)	kg	assumed 1.5 occupants with cargo = 136 kg (300 lbs. as tested in FTP cycles)
	Total Mass (2/3 tank)	kg	sum of all masses on board with 2/3 full fuel tank
Vehicle	Rolling Resistance Coeff.	—	assumed constant, = 0.009 for current, 0.008 for evolutionary, 0.006 for advanced.
	Drag Coefficient	—	assumed constant, = 0.33 for current, 0.27 for evolutionary, 0.22 for advanced.
	Frontal Area	m2	assumed constant, = 2.0 for current, 1.8 for future.
	Auxiliary Power	W	assumed constant, = 1000 W during future vehicle operation, = 700 for current vehicles.
	Power:Weight Ratio	W/kg	maximum total power available / total mass, held constant at 75 W/kg.
Engine	Engine Displacement	cm3	chosen according to engine power desired.
	Transmission Efficiency	—	assumed constant, = 0.7 for current city automatic, 0.8 for current highway automatic, 0.88 for automatice clutch and continuously variable.
	Indicated Efficiency	—	assumed constant, = 0.38 for current gasoline, 0.41 for future gasoline, 0.44 for future natural gas, and 0.51 for future diesel.
	Frictional MEP ressure	kPa	assumed constant, = 165 kPa for current gasoline, 124 kPa for future gasoline, and 153 kPa for future diesel.
	Max Engine Power	kW	maximum power from combustion engine.
Motor	Hybrid Threshold	kW	power below which hybrids are only driven with batteries.
	Gear Efficiency	kW	modeling result, dependent on load and speed.
	Electric Motor Efficiency	—	modeling result, dependent on load and speed.
	Max Motor Power	kW	maximum power from electric motor.
Fuel Cell	H2 Flow Concentration	%	hydrogen concentration available to fuel cell; affects stack efficiency.
	Fuel Cell System Efficiency	—	modeling result based on energy produced by fuel cell for road use / energy in hydrogen into fuel cell.
	Reformer & Utilization Eff.	—	energy in hydrogen consumable by fuel cell / energy stored in fuel for conversion.
	Peak Stack Power	kW	maximum power from fuel cell stack, contributing 85% of fuel cell hybrid available power.
Fuel	Lower Heating Value	MJ/kg	constants; usual to define ICE efficiency with lower heating value.
	Fuel Density	kg/L	constants.
Battery	Battery Discharge Efficiency	—	assumed constant, = 95 %.
	Specific Energy	Wh/kg	= 50 Wh/kg for hybrids, US Advance Battery Consortium commercial goal 150 Wh/kg for EV.
	Specific Power	W/kg	= 800 W/kg for hybrides, US Advance Battery Consortium commercial goal 300 W/kg for EV.
	RESULTS		
City	Fuel Energy Use	MJ/km	modeling result.
	Battery Status	MJ/km	modeling result.
	Combined Energy Use	MJ/km	vehicle energy use, specific to city driving cycle; for hybrids, battery use is adjusted by a factor to take into account final battery SOC.
	Combined Fuel Consumption	L/100km	fuel-specific consumption based on combined energy use.
	Combined Fuel Economy	mpg	fuel-specific economy based on combined energy use.
	Range (fuel only)	km	driving range of vehicle based on fuel on board and the city driving cycle, (excludes battery charge depletion at low speeds).
	Tank-to-Wheel Efficiency	%	energy supplied to wheels / total energy use; note regenerated energy not included.
Highway	Fuel Energy Use	MJ/km	modeling result.
	Battery Status	MJ/km	modeling result.
	Combined Energy Use	MJ/km	vehicle energy use, specific to highway driving cycle; for hybrids, battery use is adjusted by a factor to take into account final battery SOC.
	Combined Fuel Consumption	L/100km	fuel-specific consumption based on combined energy use.
	Combined Fuel Economy	mpg	fuel-specific economy based on combined energy use.
	Range (fuel only)	km	driving range of vehicle based on fuel on board and the highway driving cycle, (excludes battery charge depletion at low speeds).
	Tank-to-Wheel Efficiency	%	energy supplied to wheels / total energy use; not regenerated energy not included.
combined	Equivalent Energy Use	MJ/km	combined vehicle cycle energy use, with 55% city and 45% highway operation.
	Gasoline Eq. Consumption	L/100km	total energy use converted to equivalent gasoline fuel consumption.
	Gasoline Eq. Economy	mpg	total energy use converted to equivalent gasoline fuel economy.
	Cycle Carbon Emission	g C /km	carbon emitted during combined vehicle cycle.

Table 6. Brief comments on variables listed

Note that the numerical values in the summary table, which are given to several significant figures to match with the assumptions made and input variables chosen, do not have that level of precision. Predictions for 20 years into the future obviously depend strongly on the assumptions and input variables and are unlikely to be better than ±15-20% in a relative technology to technology comparison sense. Note that all columns show a substantial reduction in energy consumption and CO_2 emissions as compared to the baseline vehicle, because of the compounding effects of reducing vehicle resistances and the corresponding maximum propulsion system power, and of improving the propulsion system efficiency (both power unit and transmission).

Battery Projection

Battery technology has a limited effect to the success of ICE and FC hybrids at reducing energy consumption. Using optimistic batteries (2000 W/kg, 80 Wh/kg) instead of conservative ones (800 W/kg, 50 Wh/kg) saves about 2-4% in mass, and reduces energy consumption by only 1-2%. More important for HEV batteries will be the development in specific energy, which would extend the full-power capability of hybrids.

However, battery technology will determine the fate of the pure battery electric vehicle. Realizing the USABC commercial goal of 150 Wh/kg and 300 W/kg will enable significant energy reductions. However, if this goal is not reached, then EV range will be significantly effected.

To compare our results with calculations using a more conservative battery technology, we chose the USABC short-term desired goal of 100 Wh/kg and 200 W/kg, which some marketable Lithium ion batteries are rapidly approaching today. This more conservative battery performance increases the overall EV mass because a heavier battery pack, stronger motor, and greater structural support (with compounding effects) are needed to maintain the constant power-to-mass ratio. Already deficient in range because of the low energy density of batteries, the EV increase in mass by 34% and in fuel consumption by 18%, making the EV even less attractive.

Vehicle Load Sensitivity

Analyses for all technologies were performed with a passenger/cargo load of 136 kg (300 lbs); and to determine the consumption sensitivity to increased load, the vehicles were tested again, this time with 4 adults at 75 kg each and 20 kg of cargo, totaling 320 kg. The advanced vehicles have less mass compared to the conventional or evolutionary cars, hence a heavier passenger/cargo load would represent a larger fraction of the total moving mass, affecting the power availability and fuel consumption.

The ICE (including hybrids) vehicles show themselves to be more robust with various loading, with a larger percent mass increase (14.9-16.2%) but a smaller percent energy consumption increase (3.8-5.7%).

Meanwhile, the electric drive vehicles, including all FC models and the EV, had greater percent consumption increase (6.6-7.6%) with smaller percent mass increase (12.6-14.0%). This trend can be attributed to the nonlinear efficiency of the ICE, which can generate more power with less additional fuel use, whereas the electric motor and corresponding power units have more constant power delivery efficiencies.

Estimated Retail Price

A price summary table relative to the vehicle cycle only, showing gasoline-equivalent energy consumption and carbon emission is presented in Table 7. From the vehicle cycle, the hydrogen fuel cell and the electric vehicles offer zero-emission options at a cost, with the next best options being the diesel and gasoline ICE hybrids. Of course, the fuel cycle emissions for electricity and hydrogen must be taken into account for a total system energy and CO_2 emissions assessment. This is done in our full study report [14]. These results suggest that the methanol and gasoline reformer fuel cells, aside from research value, offer little carbon benefits at a high cost.

VEHICLE TECHNOLOGY DISCUSSION

The vehicle simulation suggest that impressive fuel economy and CO_2 emissions benefits may be realizable, with the potential scope dependent upon the assumptions made about the component performance characteristics, their subsequent compounding effect with each other, and the extent to which these new and improved technology vehicles prove marketable.

Our assumptions and results are projections of what potentially practicable vehicle and propulsion system improvements might produce in terms of reduced average passenger car energy consumption and CO_2 emissions by about 2020, with other vehicle performance attributes roughly held at today's levels. Driveability issues such as ease of start up, driving smoothness, and operation temperature range are assumed to be solvable for each technology. Transient response for rapid accelerations is expected to be addressed by sophisticated power control electronics for hybrids and short-term reserve storage for fuel cell vehicles. Power-related issues such as hill climbing, and load carrying/pulling capacity are assumed to be satisfactory given the more than adequate preset power-to-weight ratio for all technologies.

Date		1996	2020	2020	2020	2020	2020	2020	2020	2020	2020	2020
Technology		current	baseline	advanced	advanced	advanced	advanced	advanced	advanced	advanced	advanced	advanced
Propulsion System		SI ICE	SI ICE	SI ICE	CI ICE	SI Hybrid	CI Hybrid	SI Hybrid	FC Hybrid	FC Hybrid	FC Hybrid	Electric
Fuel		gasoline	gasoline	gasoline	diesel	gasoline	diesel	CNG	gasoline	methanol	hydrogen	electricity
Transmission		auto	auto-clutch	auto-clutch	auto-clutch	CVT	CVT	CVT	direct	direct	direct	direct
Gasoline Eq. Consumption	L/100km	8.46	5.44	4.79	4.20	3.32	2.86	3.20	5.56	4.14	2.50	1.58
normalized over evolutionary	%		100%	88%	77%	61%	53%	59%	102%	76%	46%	29%
Vehicle Cycle C Emission	g C /km	53.3	34.3	30.2	28.2	20.9	19.2	15.5	35.0	24.9	0.0	0.0
normalized over evolutionary	%		100%	88%	82%	61%	56%	45%	102%	72%	0%	0%
Vehicle Price	1997 US$	17200	18000	19400	20500	23300	24300		28200	27500	25500	26100
Operating Cost	US$/km	0.306	0.304	0.320	0.329	0.363	0.373		0.430	0.417	0.394	0.394
normalized over evolutionary	%		100%	108%	114%	129%	135%	0%	157%	153%	142%	145%

Table 7. Price summary relative to carbon emission

The calculations involve many numerical inputs, resulting in a certain level of uncertainty. We have attempted to be as internally consistent with these inputs as is feasible. The uncertainties are significantly less where we are extrapolating from the performance of well established technologies (such as steel chassis and body components, and spark-ignition engines). The uncertainties in performance, weight and cost, increase for technologies that have come into production relatively recently, but whose ultimate potential is still being explored (e.g. extensive use of aluminum, small low-emissions diesel engines, continuously variable transmissions). And the uncertainties become much greater for new technologies such fuel cells and high performance batteries, where the performance and cost of current versions of these technologies fall far short of what would be required for market feasibility. Here we have used literature assessments of the future development potential, tempered by our own judgments of plausible long-term technology improvements.

Technology Highlights

The advanced vehicle bodies with reduced mass raise several safety and handling issues; safety performance in required government tests must be maintained or surpassed, possibly by adding extra features to compensate for reduced energy absorption as the body is crushed. Innovations regarding safety may require extra cost and mass that we may have not included here. The customers' response to these weight reductions is unknown.

The internal combustion engine has made a place for itself in history, and will likely stay for decades to come as the most economic choice for power units. The gasoline engine remains the most used, most robust, and most well-researched of the ICE's, with the potential for further improvement in both fuel economy and pollutant reduction. Meanwhile, the diesel engine is more efficient, and has a higher power density when turbo-charged, and as a consequence is widely used for long range transportation trucks. However, its pollutant emissions, especially NOx and small particulates, given the structures of future regulations, will require the development of effective diesel exhaust treatment technologies. Innovations in exhaust treatment, even at the expense of sacrificing efficiency, will prove to be very important, particularly in Europe, where diesels are used much more extensively. Finally, compressed natural gas engines are also an area of interest for urban use because of its cleaner properties.

Hybrid technology holds much promise as the next logical step to maximize the inherent advantages of the ICE, at higher loads where it operates more efficiently. New production models are now being tested on market, with many more companies following suit. Their power control methods and technologies are more complex than the one simulated here, which is either-or, or both, depending on power requirement. However, much more can be accomplished with more sophisticated controls, not just increasing mileage, but also lowering emissions and improving performance.

Fuel cell technology is still relatively new to the automotive application, and will require much research and developing time and money. From our study here, it seems that reformer technologies are not that promising

in terms of reducing carbon emissions, while the direct hydrogen fuel cell, without the reformer losses and weight, is the one technology with the potential for adequate range and zero vehicle emissions. However, the anticipated method of production of hydrogen from natural gas has significant energy losses and CO_2 emissions, and the lack of a hydrogen infrastructure and the safety issues around hydrogen may prove to be substantial barriers to market entry. Furthermore, while mass was estimated in this study, the propulsion system volume and its direct impact on the vehicle were not. The extra space that may be needed with alternative propulsion system technologies could translate into a larger or differently shaped vehicle and increase resistances, indicating that our results are likely an upper bound for its potential performance. One of the most critical space and mass issue for the hydrogen fuel cell is the hydrogen storage, with current options being a standard compressed gas tank and a chemical hydride storage, both of which cannot store more hydrogen than 5% of the storage unit mass.

All vehicles with electric drive including the ICE and FC hybrids, and particularly the electric vehicle, are directly effected by improvements in electrical components such as more efficient motors, inverters, power electronics, and especially batteries. We have also pointed out that EV battery assumptions have drastic effects on vehicle specifications and performance.

Finally, it is worth mentioning that the EV has the potential to be a total zero-emission vehicle, if green electricity is used to recharge the batteries. All vehicles with batteries can charge from a standard outlet, which can be environmentally beneficial if the electricity source is renewable. They can also bypass the grid and charge directly from solar arrays, either mounted permanently on the vehicle, unveiled only during parked situations, or at specially designed parking spaces, thus making the electric drive even more attractive in terms of consuming sustainable, non-polluting energy, in addition to less tailpipe emissions in urban areas.

Vehicle Cycle Summary

In assessing the potential of these different technology combinations, we have ensured comparability among the vehicles by setting the power-to-mass ratio, travel range, and general size to the same levels as today. However, as researchers at Carnegie Mellon argue [15], consumers may perceive attribute "bundles" to be more flexible; that is, they may be value one feature over another for particular applications. For example, a consumer may prefer a small EV with limited range for her daily commute instead of a mid-sized EV loaded with batteries for increased range she does not need. Also, in the recent past, customers have offset vehicle efficiency gains by choosing larger and, hence, heavier vehicles. In reviewing the results of this study, one has to remember the initial assumptions that derived the conclusions.

Numerically, the results have engineering uncertainties in addition to real life driving uncertainties, and should be interpreted with these uncertainties in mind. Perhaps most useful is comparing these different vehicle and propulsion system combinations in terms of their percentage reduction in fuel consumption relative to the evolving baseline vehicle level in 2020. It is also important that the total impact of these various technologies be assessed in the context of the total fuel supply, vehicle production and recycling, and vehicle use cycle, summed together to compare the total energy consumption and carbon dioxide emission.

Finally, as explained previously, prior technology assessment literature has often underestimated the steady improvement of the baseline or mainstream technology with time, and overestimated the performance of new technology, often not considering the more pragmatic, often difficult to quantify, but important attributes (such as start-up time and refueling ease) crucial to vehicle users. Also, the time required to develop new automotive technologies to the point where they have market potential, and to design mass-production feasible versions of these technologies, has consistently been too optimistic (OTA report [16]). In our own studies reported here, we have attempted to be balanced in our assessments of established and new technologies. Our specific summary conclusions on relative vehicle technology performance and price follows; comparisons of energy consumption or fuel economy refer to fuel loaded on board the vehicle and **not** to the total well-to-wheels cycle:

1. The projected evolving baseline passenger car improvements, which are likely to be driven by market pressures and some tightening of regulations, are significant: 15% reduction in vehicle mass, 35% reduction in fuel consumption, and about 5% price increase, as compared to today's average car.

2. The more advanced vehicle technology car, with the same improved baseline gasoline engine and improved transmission, and with further reductions in mass and resistances, decreases the mass by an additional 8% and fuel consumption by a further 12%, relative to the 2020 evolving baseline car, at an additional 8% price increase.

3. The gasoline hybrid with its CVT shows a further reduction of about 30% compared to the advanced gasoline ICE vehicle, and almost 40% compared to the evolving car, but at a price increase of 20% and 30% relative to the advanced and evolving vehicles respectively. Note also that the hybrid has a significant advantage during the urban driving cycle.

4. The compressed natural gas hybrid is only slightly more fuel conserving than the gasoline hybrid; however, it has a 25% reduction in carbon dioxide emission. Its greenhouse advantage does not include methane venting as is likely to happen during CNG distribution and refueling.

5. The diesel engine provides extra energy advantages at an increased initial cost and with possible emissions compliance issues. The advanced diesel ICE vehicle is about 12% less energy consuming but some 5% more expensive than the gasoline version, while the diesel hybrid is roughly 15% less energy consuming and also about 5% more expensive than the gasoline hybrid.

6. The most efficient and lowest emitting vehicle of the group is the direct hydrogen fuel cell hybrid vehicle, with nearly 55% lower energy consumption than the evolving ICE vehicle, at about a 40% price increase. However, this assessment only accounts for the vehicle cycle, which does not include the energy and

emissions associated with manufacturing and distribution of hydrogen.

7. The reformer fuel cell vehicles, which process liquid fuels to hydrogen on board, perform significantly worse than the hydrogen fuel cell and have the highest prices of the group studied, with the gasoline reformer consuming more fuel, emitting more carbon, and having a higher cost than the methanol version. Overall, the gasoline and methanol reformer FC vehicles uses 120% and 65% more energy than the hydrogen FC vehicle, and about 2% more and 25% less energy respectively than the evolving baseline gasoline ICE vehicle. At a cost over 55% and 50% more expensive than the evolving baseline for the gasoline and methanol reformer FC vehicles, these technologies appear unattractive.

8. While the battery electric propulsion system requires the lowest energy input (as electricity) to the vehicle, with either the optimistic or more conservative assumptions about future battery technology, when allowance is made for the efficiency of electricity production and distribution, the total energy consumed is dramatically higher. Hence, it is not appropriate to compare the EV to other technologies based only on the vehicle cycle. Also, given the battery technology considered, the EV is about 45% higher in price than the baseline.

ACKNOWLEDGEMENT

The analyses described here is part of a larger fuel and vehicle technology project. The authors acknowledge the assistance and helpful discussions with Dr. Malcolm Weiss, Dr. Elisabeth Drake, and Darian Unger. This project was supported by the MIT Consortium on Environmental Challenges (CEC) and by Aramco, Chevron, Exxon, and Mobil. The CEC is an industrial-governmental-academic-foundation partnership to explore the role of science, technology, and social science in the formation of better environmental policy; it is funded by the Ford Motor Company, Norsk Hydro, and Exxon.

CONTACT

Professor John Heywood can be reached through the Department of Mechanical Engineering, Massachusetts Institute of Technology at (617) 253-2243, or email jheywood@mit.edu. Author Felix AuYeung would be happy to discuss all issues, technical or social, and can be reached by email at auyeung@alum.mit.edu.

REFERENCES

1. "World Total Energy Consumption," United States Department of Energy, <www.eia.doe.gov/oiaf/ieo95/tbls.html>.
2. Ward's Automotive Yearbook 1997, 59th Edition, Ward's Communications, 1997.
3. Motor Vehicle Facts & Figures 1998, American Automobile Manufacturers Association, 1998.
4. Dale Chon and John Heywood. "Performance Scaling of Spark-Ignition Engines: Correlation and Historical Analysis of Production Engine Data," SAE Paper 2000-01-0565, 2000.
5. L. Guzzella, A. Amstutz. "Quasi-Stationären-Simulations," Matlab programs and text Benutzeranleitung, Laboratorium für Motorsysteme, Institut für Energietechnik, ETH-Zürich, 1998.
6. EPA annual Fuel Economy Guide. <www.epa.gov/omswww/mpg.htm>, 2000.
7. John B. Heywood. Internal Combustion Engine Fundamentals, McGraw-Hill, Inc., 1998.
8. GM EV1- Specifications. <www.gm.com>, 2000.
9. "Hybrid Electric Vehicle Component Information" Department of Energy, <http://www.hev.doe.gov/components/batteries.html>, August, 1999.
10. United States Advanced Battery Consortium Battery Criteria: Commercialization, <www.USCAR.org>, May, 2000.
11. Fritz Kalhammer.
12. David J. Friedman. "Maximizing Direct-Hydrogen PEM Fuel Cell Vehicle Efficiency – Is Hybridization Necessary?" Society of Automotive Engineers, 1999-01-0539, 1999.
13. C.E. Thomas, et al. "Societal Impacts of Fuel Options for Fuel Cell Vehicles," Society of Automotive Engineers, 982496, 1998.
14. M.A. Weiss, J.B. Heywood, E.M. Drake, A. Schafer, F. AuYeung, D. Unger. "On the Road in 2020: A Life-Cycle Analysis of New Automobile Technologies," Energy Laboratory Report, M.I.T., October, 2000.
15. H.L. MacLean, L.B. Lave. "Addressing Vehicle Equivalency to Facilitate Meaningful Automobile Comparisons," Society of Automotive Engineers, 1998.
16. Office of Technology Assessment. Advanced Automotive Technologies: Visions of a Super-Efficient Family Car, OTA-ETI-638, Washington D.C., 1995.
17. Michael Kluger and Denis Long. "An Overview of Current Automatic, Manual and Continuously Variable Transmission Efficiencies and their Projected Future Improvements," SAE Paper 1999-01-1259, 1999.
18. "Issues in Midterm Analysis and Forecasting 1999," Energy Information Administration, Department of Energy, 1999.
19. "NAIAS Highlights: Concepts," Automotive Engineering International, March 2000.
20. "New Ford Hybrid Electric Car Tops 70 mpg," Ford Motor Company: Better Ideas: Environment, <www.ford.com>, 2000.
21. A.F. Burke. "Hybrid/Electric Vehicle Design Options and Evaluations," Society of Automotive Engineers, 920447, 1992.
22. Bosch Automotive Handbook, 4th Edition, Robert Bosch GmbH, 1996.
23. Dieter Stock and Richard Bauder. "The New Audi 5-Cylinder Turbo Diesel Engine: The First Passenger Car Diesel Engine with Second Generation Direct Injection," Society of Automotive Engineers, 900648, 1990.
24. Greg Ayres. "Consumer Incentives to Reduce Greenhouse Gas Emissions from Personal Automobiles," Society of Automotive Engineers, 1999-01-1307, 1999.
25. Hartmut Lüders, Peter Stommel, Sam Geckler. "Diesel Exhaust Treatment- New Approaches to Ultra Low Emission Diesel Vehicles," Society of Automotive Engineers, Society of Automotive Engineers, 1999-01-0108, 1999.
26. Honda Insight, <www.honda.com>, 2000.
27. James Adams, et al. "The Development of Ford's P2000 Fuel Cell Vehicle," Society of Automotive Engineers, 2000-01-1061, 2000.
28. P.J. Shayler, J.P. Chick, D. Eade. "A Method of Predicting Brake Specific Fuel Consumption Maps," Society of Automotive Engineers, 1999-01-0556, 1999.
29. Richard K. Stobart. "Fuel Cell Power for Passenger Cars-What Barriers Remain?," Society of Automotive Engineers, 1999-01-0321, 1999.
30. Toyota- Prius- Specifications. <toyota.com>, May, 2000.
31. William L. Mitchell, M. Hagan, S.K. Parbhu. "Gasoline Fuel Cell Power Systems for Transportation Applications: A Bridge to the Future of Energy," Society of Automotive Engineers, 1999-01-0535, 1999.
32. "Detroit Auto Show: Ballard Announces Production-Ready Fuel Cell Module, Hints at Factory Plans," The Hydrogen & Fuel Cell Letter, ISSN 1080-8019, 2000.
33. "Detroit Auto Show: GM Rolls out Fuel Cell "Precept" Concept, Claims 500 Mile Range," The Hydrogen & Fuel Cell Letter, ISSN 1080-8019, 2000.
34. "Fuel Squeeze Drives Interest in Concept Cars," Environmental News Service, 2000.
35. "Electric Assist," Electomotive, Inc., 9131 Centreville Road, Manassas, VA 02110.

Managing Regulatory Content

Monica H. Prokopyshen and Ross G. Good
DaimlerChrysler Corp.

ABSTRACT

Managing regulatory content is a complex process for any industry, but particularly for the automotive industry, which is heavily regulated. Several approaches for managing content are discussed along with implications for the industry. The response of an Original Equipment Manufacturer (OEM) to the recent European Parliament End-of-Life Vehicle Directive (EU 2000/53/EC-ELVs)[1] is discussed from a North American perspective as well as trickle down expectations for the automotive supply base. Design, sourcing and labeling issues associated with the ELV directive as well as domestic regulations are discussed.

INTRODUCTION

Managing regulatory content is a complex process for any industry, but particularly for the automotive industry, which is heavily regulated. Robert Samuelson of Newsweek provided the following U.S. perspective on regulations. There are 202 volumes containing 32,000 pages, which ultimately cost consumers $500 billion annually.[2] In addition to a growth in regulations, there is the increasing risk of civil and criminal prosecution and fines. Both the number of substances regulated and the number of laws continue to increase.

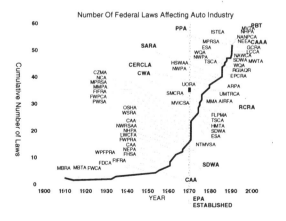

SUBSTANCES OF CONCERN - There are a hundred thousand or so chemicals in commercial use today. Of these, over a thousand chemicals have been classified as substances of concern in the United States with attendant reporting, tracking and limits on use and exposure. Requirements may be imposed at the municipal, county, state or federal level by various agencies and legislative and regulatory bodies. The same chemicals may be found in different laws or regulations yet have different actionable levels. The substances targeted and the concentrations of concern are as varied as the number of agencies, laws and regulations that spawned them. At the federal level, of the 50 titles of the United States Code (Federal Statutes), ten are the main sources of substance tracking, monitoring, labeling, classification, communication and identification requirements for industry:

14 Coast Guard
  (water transport of hazardous materials)
15 Commerce and ForeignTrade
19 Customs Duties
26 Internal Revenue
29 Labor (OSHA)
30 Mineral Resources
33 Navigation and Navigable Waters
40 Protection of Environment
43 Public Lands: Interior
49 Transportation
  (Marking, labels, packaging, classification of hazardous materials).

The list of federal regulations and laws grows very rapidly within each of these sections. Appendix "A" provides a sample of federal laws related to hazardous materials or substance management. State laws address many of these same topics and are on average more stringent than federal requirements. Appendix "B" provides a sample of Michigan environmental laws that OEMs need to consider.

CONTENT DEPENDENT REQUIREMENTS – This paper provides a few examples of content dependent requirements to illustrate the varied sources and complexity of managing this issue.

Ozone Depleting Substances One requirement is a result of the 1987 Montreal Protocol on Substances that Deplete the Ozone Layer. A federal excise tax is imposed on parts imported to the United States in which ozone depleting substances were used during manufacture. There are both fixed and unit of production cost aspects associated with this tax. The tax imposes a requirement on the OEM to track the use of ozone depleting substances utilized in the manufacture of supplier parts. Note that the definition refers to "used in the manufacture", not necessarily contained in the component and the scope extends beyond the borders of the United States as well as outside the OEM organization.

Federal Procurement Guidelines The second example illustrates the indirect effect of a federal procurement requirement. In 1995, the Environmental Protection Agency released for comment "Guidance on Acquisition of Environmentally Preferable Products and Services,"[3] in response to Executive Order 12873 entitled "Federal Acquisition, Recycling and Waste Prevention." As this rolled out, States like Pennsylvania began evaluating bids for vehicle fleet purchases by considering the recycled content of vehicles, on a component by component basis! Up until this point the discussion focussed on substances of concern, however, this example introduced the need to manage recycled content as well. Prior to this point, manufacturers certified the recycled content of a product as a whole—now manufacturers were asked to provide detailed component information.

Government agencies were not the only customers who requested product environmental performance information in advance of purchase. In 1995, commercial fleet customers like Northern Telecom began to ask about recycled content and environmental programs to gauge the environmental performance of various OEMs.

NAFTA Content Reporting The content initiative whose effects are perhaps most commonly known to the automotive industry resulted from the North American Free Trade Agreement (NAFTA). In 1993 Chrysler, General Motors and Ford allied themselves with the Automotive Industry Action Group (AIAG) and the American Automobile Manufacturers Association (AAMA) to develop a common *Supplier Content Reporting Requirements* workbook and instruction program.[4] The common approach helped the OEMs as well as suppliers manage a collective, burdensome reporting obligation.

As with many new and complex rules and regulations, there were many reactions to the prospect of NAFTA rules and regulations. They ran the gamut from "Let's prepare ourselves for the eventuality," or "Let's wait and see," to "It won't happen." We experienced a similar range of reactions within Chrysler when we forecast mercury labeling laws in the United States and the severe restriction of mercury use in automobiles sold in Sweden. Some vehicle platforms immediately initiated development programs to replace mercury in the few applications that existed at the time. One engineering group felt that lobbying efforts should be the key management focus. Still another engineering group responded: "Tell me when it becomes law." The wait was less than two years. Responses were as varied as the requirements.

As demonstrated, content requirements come in many guises from the tax code, to trade agreements to customer demands. Prohibitions, labels, "community right to know" legislation, reporting requirements, national and international material standards, third party certifications, eco-labels, "green seals", registration and facility operating permits are additional examples of the myriad origins of content management requirements. Requirements may originate from import regulations, tax law, the Occupational Health & Safety Administration, the Department of Transportation, the Environmental Protection Agency, other agencies noted earlier and standards such as the Federal Motor Vehicle Safety Standards--and this is just in the United States.

The problem compounds when the global nature of the automotive industry is considered. To put this in perspective, the DaimlerChrysler Corporation (the Chrysler Group of DaimlerChrysler AG) sells over 90 different models in over 100 countries. Of these countries, thirty are regulated markets.

So how does a company manage all these regulations, standards, substances of concern and materials?

MANAGING CONTENT

Part of the answer lies in the policies of the company and another part in the business needs for content information. Managing content is basically a knowledge management exercise. When managed in an ad hoc manner, costs and risks are increased throughout the corporation. The balance can be tipped towards benefits and lower implementation costs through a consistent, systematic corporate-wide approach. The risk of an ad hoc approach is that multiple departments may track the same regulations for differing purposes, creating process, communication and documentation inefficiencies and gaps. For example, the vehicle homologation group may track the ELV directive for the purpose of certifying vehicles for sale in Europe, whereas regulatory groups may track the directive for the purposes of identifying labeling requirements. Material engineering, environmental or recycling groups may focus on end-of-life disposal issues. With fragmented approaches, the risk also increases that suppliers will receive mixed signals on the importance and significance of specific content initiatives. These messages could be positive (reinforcing) or negative (conflicting).

Earlier, the complexity of managing content was discussed with respect to the varied sources of requirements, however complexity also increases with the number of groups who administer the rules and regulations within a company as well as the size of a company. For example, coordination in the DaimlerChrysler Corporation must occur among many groups including:

- Industrial Hygiene & Toxicology
- Environmental Groups
- Vehicle Safety and Emissions Certification Groups
- Aftermarket Sales & Marketing Groups
- Engineering Groups (Materials, Recycling, Vehicle Homologation, Releasing, Design, Management)
- Logistics & Transportation
- Production Control, Procurement & Supply Organizations
- Office of the General Counsel
- Tax Affairs
- Product Strategy Groups
- General Accounting Office (Audit Group)
- Technical Teams
- Suppliers.

Communication complexity increases exponentially with organizational size.

APPROACHES FOR MANAGING CONTENT – Six approaches are discussed.

- Wait and See
- Market Selection
- Sales Approaches
- Supplier Leadership
- Supplier Management
- Knowledge Management.

Wait and See This doesn't differ from the end-of-pipe approach seen in United States environmental regulations of the seventies and eighties. The focus was on controlling the outputs at various points in the manufacturing process (stack, waste-water discharge, solid waste disposal) rather than looking at the entire system. The problem with this approach is that it usually results in a sub-optimal solution. Although measures can be introduced that comply with the law (carbon wheel absorption units to absorb volatile organic compounds) these may involve significantly more capital and operating expenses than solutions that focus on managing the inputs or raw materials— the source of the volatile organic compounds.

In the mid-nineties, the introduction of water borne basecoat technology as a replacement for traditional solvent based paint enabled the St. Louis Assembly plant to avoid a significant capital outlay for pollution controls and avoid annual operating expenses in excess of one million dollars, while preserving manufacturing flexibility. The plant was in an ozone non-attainment zone and the new material allowed the plant to maintain production when other facilities in the city, using the traditional technology, were forced to temporarily curtail production.

An advantage of the *wait and see* approach is that resources are neither deployed too soon nor when not required. Emerging regulations may change or their introductions may be delayed and the *wait and see* approach permits a response based on a fully developed requirement.

Market Selection Another approach that may be used is to limit the markets in which vehicles are sold. Knowledge of the specific country requirements and costs of compliance provide input to the marketing, sales and product strategy organizations. It has happened that the cost of compliance exceeded the anticipated revenue from the sale of a small number of vehicles and the decision was made not to offer a specific vehicle line in a given country.

<u>Sales Approaches</u>   These may cover component or product sales and may result from regulatory drivers or internal company policy. One example is the *core program.* To minimize the risk of improper disposal, batteries are part of core programs—parts that when returned to the dealership result in a discounted price on replacements. The supply/logistics partner is the battery manufacturer, who ensures that batteries are recovered and properly recycled. Another example of a sales approach to managing content requirements is the sale of products like extended life antifreeze, designed for one hundred thousand miles, reducing the risk of improper disposal or spill during the use phase of the vehicle. In addition to extended life, the antifreeze is formulated with 10% recycled glycol content. The sales approach could be extended beyond the management of components to the management of end-of-life vehicles.

The ELV directive places a take-back burden, disposal prohibitions (depollution requirements) and landfill avoidance requirements on the OEM, "irrespective of the how the vehicle has been serviced or repaired during use and irrespective of whether it is equipped with components supplied by the producer or with other components whose fitting as spare or replacement parts accords with the appropriate Community provisions or domestic provisions."[5] Educating consumers in the environmental consequences of their use phase choices can help minimize end-of-vehicle-life consequences for the OEM. Lease programs could offer a solution to control vehicle modifications by the consumer and manage end-of-life take back requirements. This type of approach has been adopted by Xerox.

<u>Supplier Leadership</u>   In this approach, reliance is placed on the supplier to manage regulatory and customer content requirements appropriately. Requirements may be spelled out explicitly in engineering standards and contract clauses or indirectly through clauses that mandate compliance with all relevant laws and regulations. For the supplier leadership approach to be most successful, the supplier must understand and apply the same priorities, interpretation and requirements as the OEM to content reporting. This is not always the case.

For example, the Industrial Hygiene and Toxicology organization of DaimlerChrysler Corporation requests additional information beyond the basic Material Safety Data Sheet (MSDS) requirements to better manage occupational health, occupational safety and environmental protection in products and facilities.

When the Regulated Substance and Recyclability Certification (RSRC) system for vehicle components was launched by the Product Strategy and Regulatory Affairs department in 1994-95, the majority of suppliers were not tracking materials and substances of concern in articles. There were knowledge gaps as well. Electrocoat corrosion protection finishes on steels were reported by material suppliers, but not all part suppliers understood that electrocoat may contain lead.

In other cases the supplier's material knowledge is greater than that of the OEM (design and composition knowledge, country specific knowledge, etc.). Here it is important that the lines of communication between the supplier and OEM be open and measures to protect intellectual property be in place to ensure that important material and substance information is shared and acted upon.

The recent tire litigation demonstrates the risks and reputation damage that may occur in data management. The recent recall of taco shells is another example of the consequences of relying on suppliers to manage product content. In this case a genetically engineered product approved only for animal feed was introduced into products for human consumption.[6]

<u>Supplier Management</u> This approach builds upon the advantages of the supplier leadership approach, but the OEM exercises more control in managing regulations and over the process of content reporting, tracking and material management. For example, the OEM dictates the materials and substances to be tracked and interprets and manages the data. These may be weight data for the production part approval process, or material and substance information for the RSRC process. Internal and external suppliers input information to the RSRC system and these data are reviewed and acted upon. Feedback is provided to suppliers and internal customers and data are used to support a number of content dependent initiatives including material screening, identification of improvement opportunities (recyclability, recycled content, substances of concern, materials), life cycle studies, and labeling requirements.

<u>Knowledge Management</u>   This is the direction the DaimlerChrysler Corporation is heading. It is an integrated approach within the company, based on the premise that information is key. A knowledge of the requirements--regulatory, performance and customer, coupled with a thorough knowledge of product and material compositions--enables the company to address current and emerging issues. Knowing what you have is the key to knowing how to manage. Making this information available to internal customers on demand is mandatory to ensure that it is used to make more informed business decisions.

KNOWLEDGE OF REGULATIONS

Methods and tools available for tracking and managing regulations are similar to those for capturing the *voice of the customer*. Company strategies for managing content regulations may progress as follows:

- Comply with the Law
- Influence Regulations
- Improve Efficiency
- Holistic Approach
- Product Stewardship

COMPLY WITH THE LAW – Complying with the law is the starting point for any company, but the exercise of determining and keeping up with the laws is not easy. Information on new laws or changes can come from many sources:

- Regulatory and Government Affairs Offices
- Country Liaisons
- Subscription Services (Environment Watch, International Environmental Reporter, etc.)
- Public and Private databases
        (www.epa.gov, www.ihsenv.com)
- Government Publications (Federal Register, etc.)
- Industry sources and meetings.

Next, the laws must be interpreted to extract pertinent information and requirements for the business. The requirements need to be communicated to engineering, procurement, manufacturing, logistics and design groups. Quality Function Deployment (QFD) is one approach for translating the voice of regulators into product requirements. Regulations are typically communicated through:

- Engineering and material standards
- Material policies and strategies
- Design guidelines, templates and tools
- Internal databases, groupware, and web sites

The ELV directive will be used to demonstrate one approach to managing regulatory content.

ELV DIRECTIVE OVERVIEW

A new directive of the European Parliament and Council on End-of-Life Vehicles (ELV) has been ratified by the European Union. The directive has several important implications for Chrysler Group vehicles that are destined for sale in Europe. The premise of the directive is "Extended Producer Responsibility." This means that automobile manufacturers will share in the burden (or reward) of proper disposal or treatment of ELVs. The object is to prevent waste and to improve the environmental performance of all economic operators involved in the treatment of ELVs. Objectives include:

- Rigid recycling/reuse/recovery requirements
- Bans of certain heavy metals
- ELV treatment.

One of the more significant requirements of the ELV directive encourages manufacturers and suppliers to limit the use of hazardous substances in vehicles. Particular emphasis is placed on the use of lead, mercury, cadmium and hexavalent chromium. The use of these elements is prohibited unless exempted in Annex II of the ELV directive.   Annex II also includes identification and labeling provisions.

Articles 4, 6, and 8 reference earlier directives such as 67/548/EC, "Classification, Packaging and Labelling of Dangerous Substances." Article 4 enjoins manufacturers from the use of hazardous substances from vehicle conception onwards and article 8 introduces burdensome tracking of substances for the purposes of dismantling instructions.

Section 8 of the directive also requires OEM's to provide dismantling instructions to facilitate the identification of components and materials that are suitable for reuse and recovery. Instructions for each class of vehicle are required six months after the vehicle has been offered for sale in the European Community.

The ELV directive also requires producers to publish a report on the design of vehicles and their components, with a view to their recoverability and recyclability. Further, OEM's are required to make this information accessible to prospective buyers of vehicles as a part of promotional literature used in marketing.

Most of the responsibility for establishing an ELV collection system is placed on the Member States of the EU. Manufacturers may be required to provide financial assistance in developing these collection sites. Furthermore, Member States are required to develop a "de-registration system," which removes a vehicle from tax roles only after it has been disposed of in an authorized facility.

The most expensive part of the directive, from the OEM's perspective, is that the delivery and treatment of an ELV is the responsibility of the manufacturer. The entire process, from transportation to final vehicle disposition, must all be accomplished without any cost to the last owner/possessor of the vehicle - regardless of whether the manufacturer or subsequent treatment facilities incur costs.

January 1, 2006 requirements for ELV reuse and recovery are set at 85% (including an allowance of 5% for energy recovery) by average weight for vehicles produced since 1980. These targets are increased to 95% reuse and by January 1, 2015. All of these targets will be re-evaluated by 2005, with the prospect for tougher targets arising.

IMPLICATIONS OF THE ELV DIRECTIVE

The first impact of this directive is that all parts must be reviewed to determine whether the four targeted substances are present in any components and whether the exemption criteria are met when these substances are present.

If the OEM or supplier already has a system for tracking content, this is an incremental effort. For example, lead, mercury, chromium and cadmium and other substances and compounds were already tracked by the RSRC system. This provided a rich source of information on thousands of components and assemblies from over 2400 supplier locations and enabled the Chrysler Group to assess the potential impact of the ELV directive.

OEM Impact - As indicated earlier, all parts need to be reviewed. The first reason is to determine whether any of the prohibited substances are present. The second reason is to determine recyclability at the detail part level, assembly level and vehicle level and to determine components which must be removed during the *depollution* phase of ELV treatment. These components would also have to be declared and identified for registered ELV treatment facilities.

An internal study indicated that a significant number of components are impacted by the ELV substance prohibitions. Of the components impacted, almost a third were fasteners. These were of lower concern because 61% had some degree of exemption under the ELV directive. Of the remaining components, close to two-thirds had no degree of exemption as the directive is now written.

These results indicated a number of actions for sales of vehicles in Europe.

- Eliminate the substances not covered by exemptions.
- Homologate parts of concern.
- Negotiate for further exemptions
- Negotiate definitions.

Eliminate Substances of Concern - Parts containing substances of concern are reviewed to determine whether the substances can be eliminated through product or process design changes. For highly engineered assemblies and safety components, this may involve extensive development and testing to qualify new designs and a transition period for implementation. Other solutions may exist as "library" or "best practices" that can be incorporated more quickly.

Some substances of concern, like hexavalent chromium [Cr(VI)] in corrosion protection coatings, are capped at 2 g / vehicle. This introduces an additional requirement for OEMs--rolling up the amount per vehicle and determining the components for which alternatives or substitutions are not currently available. Although trade-off decisions are common in the vehicle development process, the limit on Cr(VI) is a new requirement to be managed by vehicle platforms.

Homologation - In some cases, the business decision may be to modify certain components for vehicles sold in Europe.

Negotiate for further exemptions - The list of exemptions did not consider a number of common commercial metal alloys as well as other materials which contain trace or *unintentionally added* amounts of the prohibited substances. This causes an immediate problem for certain components for which alternative designs and materials are not available. It also presents a quandary for future reuse of materials recovered from older ELVs that were introduced to the market prior to the ratification of the ELV directive.

Negotiate definitions - A list of 80 to 90 terms that need clarification and definition has been created for negotiations. Within this list are terms like *new vehicle, model, recyclability,* and *recycled material content.* These terms define how calculations will be made, reporting requirements and the timing for action.

Supplier Impact - With over 70% of the vehicle components provided by external suppliers, reporting, tracking and design change obligations will be placed on suppliers. Suppliers should also contact OEMs with recommendations of parts or materials that should be considered in discussions for exemptions to the ELV directive. Increased emphasis will be placed upon suppliers for complete, timely and accurate reporting to DaimlerChrysler Corporation's web based content reporting system, RSRC. As early as 1995, discussions with material suppliers included assessments of willingness to take back components. It is possible that take-back requirements for problematic components or materials will be extended to additional suppliers by OEMs.

CONCLUSIONS

The ELV directive has provided a regulatory requirement for eliminating lead, mercury, cadmium and Cr(VI) from automobiles sold in the European Community and for calculating and reporting on vehicle recyclability. A considerable number of components are affected by the substance prohibition, including components and materials for which design alternatives do not currently exist. The directive has placed additional design contraints upon OEMs and has called into doubt the ability to use common commercial materials, including metallic alloys and recycled materials which contain trace and *unintentionally added* substances. These materials have not been covered in the exemption list of the ELV directive. Greater emphasis will be placed on suppliers for complete, accurate and timely reporting of materials and substances of concern in vehicles to help OEMs manage regulatory content. Take back requirements imposed upon OEMs may lead to similar implications for suppliers.

A number of approaches for managing regulatory content were discussed, but holistic and knowledge management approaches are the focus of DaimlerChrysler Corporation. Knowing the regulations and content facilitates content management.

REFERENCES

[1] OJ L 269, 21.20.2000, p. 34. Directive 2000/53/EC of the European Parliament and the Council of 18 September 2000 on end-of life vehicles. (OJ is the abbreviation for Official Journal of the European Communities)
[2] Samuelson, Robert J., "The Regulatory Juggernaut," Newsweek v. 124, Nov. 7, 1994, page 43.
[3] Federal Register, Vol. 60, No. 189, Friday September 29, 1995
[4] TCPC-W Supplier Content Reporting Requirements Workbook, Version 03.00, Automotive Industry Action Group, Southfield Michigan, September 1996.
[5] Directive 2000/53/EC-ELVs Article 3(1).
[6] "Kraft Recall Highlights Safety Measures," Wall Street Journal, Tuesday September 25.

CONTACTS

Monica H.L. Prokopyshen mp5@dcx.com
Ross G. Good rgg3@dcx.com

ADDITIONAL SOURCES

www.epa.gov
EPA Common Sense Initiative Automobile Manufacturing Sector Life Cycle Management/Supplier Partnership Project

DEFINITIONS, ACRONYMS, ABBREVIATIONS

OSHA: Occupational Safety and Health Administration, also Occupational Safety and Health Act 29 CFR 1900-1910.

APPENDIX A

Clean Air Act & Ammendments (CAAA 42 USC 7401-7671q)
Clean Water Act (CWA 33 USC 1251-1387)
Chemical Weapons Convention Implementation Act
Comprehensive Environmental Response, Compensation & Liability Act (CERCLA 42 USC 9601-9675)
Consumer Product Safety Act
Emergency Planning & Community Right to Know Act (EPCRA- 42 USC 11001-11050)
Federal Food, Drug and Cosmetic Act (FFDCA 21 USC 301-397)
Federal Hazardous Substances Act
Federal Insecticide, Fungicide and Rodenticide Act (FIFRA-7 USC 136-136v)
Hazardous and Solid Waste Act (HSWA)
Hazardous Material Transportation Act (HMTA 49 USC 5101-5127)
National Environmental Policy Act (NEPA 42 USC 4321-5370e)
Occupational Health & Safety Act (OSHA 29 USC 651-678)
Oil Pollution Act (OPA 33 USC 2701-2761)
Poison Prevention Packaging Act
Pollution Prevention Act (PPA 42 USC 13101-13109)
Resource Conservation & Recovery Act (RCRA 42 USC 6901-6992k)
Safe Drinking Water Act (SDWA 42 USC 300f-300j-26)
Superfund Amendment and Reauthorization Act (SARA)
Toxic Substances Control Act (TSCA 15 USC 2601-2692)
United States Postal Service Regulations

APPENDIX B

Air Pollution Control
Battery Disposal
Environmental Response
Groundwater & Freshwater Protection
Hazardous Waste Management
Liquid Industrial Wastes
PCB Compounds

Pesticide Control
Scrap Tires
Solid Waste Management
Underground Storage Tank Regulations
Used Oil Recycling
Waste Minimization
Water Resources Protection
Wetlands Protection